Vaccines for Cancer Immunotherapy

Vaccines for Cancer Immunotherapy

An Evidence-Based Review on Current Status and Future Perspectives

Nima Rezaei

Research Center for Immunodeficiencies, Children's Medical Center,
Tehran University of Medical Sciences, Tehran, Iran
Department of Immunology, School of Medicine,
Tehran University of Medical Sciences, Tehran, Iran
Network of Immunity in Infection, Malignancy and Autoimmunity (NIIMA),
Universal Scientific Education and Research Network (USERN)
Tehran, Iran

Mahsa Keshavarz-Fathi

School of Medicine, Tehran University of Medical Sciences, Tehran, Iran
Cancer Immunology Project (CIP), Universal Scientific Education
and Research Network (USERN), Tehran, Iran
Research Center for Immunodeficiencies, Children's Medical Center,
Tehran University of Medical Sciences, Tehran, Iran

Academic Press is an imprint of Elsevier
125 London Wall, London EC2Y 5AS, United Kingdom
525 B Street, Suite 1650, San Diego, CA 92101, United States
50 Hampshire Street, 5th Floor, Cambridge, MA 02139, United States
The Boulevard, Langford Lane, Kidlington, Oxford OX5 1GB, United Kingdom

Notices

Knowledge and best practice in this field are constantly changing. As new research and experience broaden our understanding, changes in research methods, professional practices, or medical treatment may become necessary.

Practitioners and researchers must always rely on their own experience and knowledge in evaluating and using any information, methods, compounds, or experiments described herein. In using such information or methods they should be mindful of their own safety and the safety of others, including parties for whom they have a professional responsibility.

To the fullest extent of the law, neither the Publisher nor the authors, contributors, or editors, assume any liability for any injury and/or damage to persons or property as a matter of products liability, negligence or otherwise, or from any use or operation of any methods, products, instructions, or ideas contained in the material herein.

Library of Congress Cataloging-in-Publication Data
A catalog record for this book is available from the Library of Congress

British Library Cataloguing-in-Publication Data
A catalogue record for this book is available from the British Library

ISBN: 978-0-12-814039-0

For information on all Academic Press publications visit our website at https://www.elsevier.com/books-and-journals

Publisher: Andre Gerharc Wolff
Acquisition Editor: Glyn Jones
Editorial Project Manager: Sabrina Webber
Production Project Manager: Joy Christel Neumarin Honest Thangiah
Cover Designer: Matthew Limbert

Typeset by TNQ Technologies

Contents

List of Contributors

Mahsa Keshavarz-Fathi
School of Medicine, Tehran University of Medical Sciences, Tehran, Iran; Cancer Immunology Project (CIP), Universal Scientific Education and Research Network (USERN), Tehran, Iran; Research Center for Immunodeficiencies, Children's Medical Center, Tehran University of Medical Sciences, Tehran, Iran

Sepideh Razi
Cancer Immunology Project (CIP), Universal Scientific Education and Research Network (USERN), Tehran, Iran; Student Research Committee, School of Medicine, Iran University of Medical Sciences, Tehran, Iran

Nima Rezaei
Research Center for Immunodeficiencies, Children's Medical Center, Tehran University of Medical Sciences, Tehran, Iran; Department of Immunology, School of Medicine, Tehran University of Medical Sciences, Tehran, Iran; Network of Immunity in Infection, Malignancy and Autoimmunity (NIIMA), Universal Scientific Education and Research Network (USERN), Tehran, Iran

Saeed Farajzadeh Valilou
Cancer Immunology Project (CIP), Universal Scientific Education and Research Network (USERN), Tehran, Iran

Preface

A long time ago, infectious diseases were a major health problem threatening human life. This led to several epidemics and pandemics and resulted in a high mortality rate. Antibiotics and vaccines were lifesaving agents that contributed to the evolution of treatment and the prevention of infectious diseases. Nowadays, lots of infectious diseases are being bridled by the application of these approaches; therefore, the focus of health has shifted toward noninfectious diseases such as cancer. Cancer is a rebellious disease originating from self-cells, which possess some characteristics similar to those of normal cells. This can complicate targeting cancerous cells without causing severe side effects. Biologic targeted therapies were developed to yield a specific directed response against tumor cells. Because they aim to influence only cells containing specific targets, this approach is safer than that of toxic agents. Although immunotherapy can be applied in both targeted and unspecific manners, designing therapeutics able to induce immune responses, which single out tumor cells, theoretically results in yielding better outcomes indicative of efficacy and safety. As expected, practice does not always follow the theory. Therefore, a number of clinical evaluations are required to assess the efficacy and safety of the therapeutic alone or compared with other standards of care.

Vaccines, which were originally known as a preventive approach to infectious diseases, have become attractive immunotherapies for cancers in both the preventive and therapeutic settings. These modalities evoking active and specific immune responses aim to improve clinical outcomes besides enhancing immunological indicators of response. Many attempts have been made to assess vaccines for cancer and improve their efficacy. Over the years, cancer vaccines have had several ups and downs. Some cases of success and some of failure were recorded in their history, leading to lessons on optimizing the modality and picking the best population and combination. Vaccines are attractive therapeutics because of promising results obtained with certain vaccines in certain cancers. Advances in genomic technologies have also promoted the status of personalized vaccines for cancers and have yielded positive clinical outcomes with vaccines alone and combined with immune checkpoint blockers. The field of cancer vaccines is moving forward; a look at the background, history, and current state, which are provided in this book, assists in the development of vaccines and provides future perspectives for this modality. In this book, we first explain the immunology of cancer and immunotherapeutics applied and approved for cancer (Chapters 1 and 2). Afterwards, types of vaccines, adjuvants, and delivery systems (Chapter 3) are examined. Next, tumor antigens as targets for vaccine therapy (Chapter 4) and approaches implemented to design autologous, allogeneic and personalized vaccines (Chapters 5 and 6) are reviewed. The following sections provide various types of therapeutic vaccines, their clinical applications and efficacy (Chapters 7–10), examples of vaccines for various types of cancers and candidates for vaccine therapy (Chapter 11), hurdles in the way of cancer vaccine development (Chapter 12), combination therapy (Chapter 13), and concluding remarks and future perspectives (Chapter 14).

We hope that this book will be welcomed not only by clinicians but also by basic scientists who wish to have an update in this field.

Nima Rezaei, MD, PhD
rezaei_nima@tums.ac.ir
Mahsa Keshavarz-Fathi, MD
m-keshavarz@student.tums.ac.ir

CHAPTER

CANCER IMMUNOLOGY

1

Mahsa Keshavarz-Fathi[1,2,3], **Nima Rezaei**[3,4,5]

School of Medicine, Tehran University of Medical Sciences, Tehran, Iran[1]*; Cancer Immunology Project (CIP), Universal Scientific Education and Research Network (USERN), Tehran, Iran*[2]*; Research Center for Immunodeficiencies, Children's Medical Center, Tehran University of Medical Sciences, Tehran, Iran*[3]*; Department of Immunology, School of Medicine, Tehran University of Medical Sciences, Tehran, Iran*[4]*; Network of Immunity in Infection, Malignancy and Autoimmunity (NIIMA), Universal Scientific Education and Research Network (USERN), Tehran, Iran*[5]

INNATE AND ADOPTIVE IMMUNITY

Pathogens and endogenous dangerous mutated and cancerous cells must be distinguished and destroyed by the immune system, which has two main arms: innate and adaptive immunity. Each arm has its own specialized cell-based and humoral responses. The first responder to exogenous and endogenous threats is the innate immune system, which operates as a nonspecific arm and rapidly acts through pattern recognition receptors (PRRs). These receptors are located on the surface of innate cells including tissue-resident cells such as macrophages, dendritic cells (DCs), monocytes, and neutrophils, which circulate in the blood. Most of the PRRs bind to the pathogen-associated molecular patterns (PAMPs) and damage-associated molecular patterns (DAMPs) to recognize the potential danger.[1,2] Toll-like receptors (TLRs) are one of the significant PRRs, present on the surface of antigen-presenting cells (APCs), such as DCs. By recognition of PAMPs and DAMPs, they initiate activation of the signaling pathways such as transcription factor nuclear factor-kappa B (NF-κB) and interferon regulatory transcription factor, which are inflammatory and induce type I interferons and cytokines to recruit and activate lymphocytes.[3,4]

The innate immune system is not capable of forming an immunological memory. Therefore, the role of the adaptive immune system in providing immunological memory and specific responses is manifest. Adaptive immune cells are capable of recognizing a single specific antigen because each lymphocyte, before facing any antigen, carries only one receptor, which is specific for one antigen. Therefore, to cover recognition of the variety of antigens, which the immune system meets through its lifespan, millions of antigen-specific receptors must exist. In order to provide this vast variety, the genes of variable chains of the receptors randomly recombine during development of lymphocytes in the central lymphoid organs, the bone marrow, and thymus. Then, various variable chains are paired to create the whole lymphocyte receptor repertoire of a person.[5]

Clonal selection of lymphocytes is the central feature of the adaptive immune system, which leads to developing specific responses. As described earlier, lymphocytes bear a variety of receptors specific

Vaccines for Cancer Immunotherapy. https://doi.org/10.1016/B978-0-12-814039-0.00001-1

for different antigens, named antigen-specific receptors. These cells are activated and proliferated only after exposure to the specific antigens. The receptor of a lymphocyte's descendants, i.e., the effector cells, are the same with their ancestor's, and this is the concept of a clone. Clonal deletion is also crucial to omit autoreactive lymphocytes, which respond to the self-antigens.[5]

To induce an immune response against cancer, two central phases are performed, i.e., the priming and effector phases. In the first phase, the APCs such as DCs prime the T cells. They obtain the tumor antigens of dying cancer cells. If a danger signal is not available, immune tolerance toward the antigen is induced. Immunogenic cell death is responsible to generate danger signals, which are recognized by PRRs on DCs. The stress induced during cell death leads to providing danger signals such as type I interferons, chemokine ligand 10 (CXCL10), CXC-chemokine receptor 3 (CXCR3), heat shock protein 70 kDa (HSP70), and HSP90. The danger signals function as adjuvants to increase the immunogenicity.[6] Antigens and danger signals lead to the maturation of DCs, and then they travel toward the draining lymph nodes.[6] To induce effector T cells, DCs transduce three signals to T cells. The first signal is transduced through antigen presentation by the major histocompatibility (MHC) molecules on DCs to T cell receptors (TCRs) on the T cells. The second signal results from costimulatory or coinhibitory molecules. The second signal adjusts and modulates the type of immune response against the danger signal. Following these signals, in the third signal a number of cytokines are produced to direct the type of following immune response.[7] The type of DC maturation affects on determining the phenotype of T cells (Fig. 1.1). As a consequence of priming, CD4+ and CD8+ effector T cells, necessary to evoke a robust immune response, are developed. Cytotoxic T lymphocytes, which are CD8+ T cells, are the main effector cells to destroy the tumor cells. However, CD4+ T cells are required for optimal and long-lived effector CD8+ T cells and for induction and maintenance of CD8+ memory cells.[8,9]

Antigens are processed through two different mechanisms. MHC-I restricted peptides undergo the proteasome dependent mechanism. The proteasome changes the long peptides to small peptides containing 9–15 amino acids, to be delivered to the endoplasmic reticulum (ER) via the transporter of antigen processing (TAP). In the ER, the peptides with 9–12 amino acids bind to the MHC-I molecules to be transported to the surface of cells. The MHC-II restricted peptides use the endosomal system. Cathepsins process the antigens in endosomes, and peptides with 12–15 amino acids bind to the MHC-II, which are transported to the surface of DCs.[10]

There are some barriers that hamper the function of effector T cells. Immunosuppressive phenotype of tumor microenvironment is one of the barriers generated due to the function of some immune cells such as myeloid-derived suppressor cells (MDSCs) and regulatory T cells (Tregs) as well as the cytokines, chemokines, and indoleamine 2,3-dioxygenase (IDO) secreted by the tumor cells.[11–14]

ACTIVATING IMMUNE CELLS

In both arms of the immune system there are immune cells and cytokines in favor of antitumor responses as well as inhibitory components, which hamper effective and robust responses against cancer. The interplay between tumor cells and these activating and inhibitory components is one of the factors determining the final dominancy of tumor cells or the immune response. Herein, we will first review the activating components and then the inhibitory building blocks of the immune response against cancer.

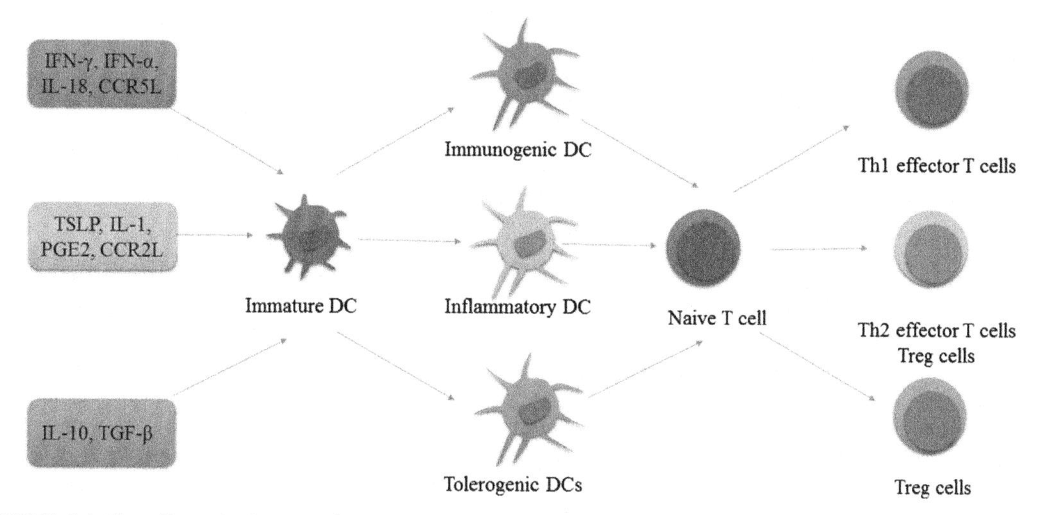

FIGURE 1.1 The effect of DC maturation on the phenotype of T cells.

Based on the various PAMPs and tissue factors, DCs may undergo one of the maturation processes including immunogenic, inflammatory, and tolerogenic maturation. Immunogenic maturation of DCs, induced by microbial components and type 1 tissue factors, which results in development of Th1 cells. Proinflammatory factors leading to inflammatory maturation of DCs give rise to Th2 and Treg cells. Both tolerogenic and immature DCs can induce Treg cells. *CCR*, CC-chemokine receptor; *DC*, dendritic cell; *IFN*, interferon; *IL*, interleukin; *PGE2*, prostaglandin E2; *TGF-β*, transforming-growth factor-β; *Th*, T helper; *Treg*, T regulatory; *TSLP*, thymic stromal lymphopoietin.

MATURE DCS

Mature DCs consist of two main subtypes, classical DCs (cDCs), also known as myeloid DCs (mDCs), and the second subtype, plasmacytoid DCs (pDCs). In the blood, a great number of DCs are pDCs, which play role in the antiviral immune response through secretion of type I interferons.[15]

There are different subsets of cDCs as well. Although CD141+ DCs are classic DCs playing an important role in cross-presentation of tumor antigens and priming of immune response against tumor,[16] all DCs, resident in lymphoid organs, such as CD141+/BDCA3+, CD1c+/BDCA1+, or plasmacytoid can do cross-presentation.[17] CD1a+ DCs, which are the major subsets of cDCs, and CD14+ DCs, with the monocyte-related phenotype, are present in the dermis layer of the human skin, whereas another self-renewing subtype of DCs known as Langerhans cell (LC) exists in the epidermis.[18] CD14+ DCs are responsible for the induction of humoral immunity, while LCs are responsible for priming of CD8+ T cells, and CD1a+ DCs perform this action as well but they are less potent in comparison to LCs.[19]

T cell priming function of DCs is initiated with presentation of MHC-peptide complexes to DCs. Costimulatory molecules on APCs and activating cytokines by T cells are subsequently produced.[20] The mentioned features determine the maturation of DCs and their direction toward antitumor immunity.

Several cytokines are produced by DCs as well. IFNs, tumor necrosis factor-α (TNF-α), interleukin 1 (IL-1), IL-6, IL-12, and IL-23 are among the cytokines secreted by DCs, which regulate T cell responses.[21] Transforming growth factor β (TGFβ), and IL-10, secreted by immature DCs, demonstrate an inhibitory effect on T cell responses and thus prevent tumor destruction.[22]

CLASSICALLY ACTIVATED MACROPHAGES (M1)

In the tumor microenvironment, macrophages can exert both anti- and protumoral roles. Their function is dependent on their phenotype developed by the effects of intracellular interplays and two main sets of cytokines.[23] Macrophages originate from immature myeloid precursors or circulating monocytes recruited into the tumor site through chemokines such as CCL2, CCL5, and CXCL12.[24,25]

Classically activated macrophages or M1 phenotype are developed under influence of cytokines such as GM-CSF, IFN-γ, and TLR agonists.[26] They play antitumoral and inflammatory roles in the tumor microenvironment by various means, including phagocytosis, antigen presentation, production of proinflammatory cytokines, and cytotoxic effects. They use reactive oxygen species (ROS), reactive nitrogen species (RNS), IL-1β and TNF-α to perform cytotoxic activities.[27] M1 macrophages produce IL-12, which promotes antitumor responses of natural killer (NK) cells and T cells by production of IFN-γ. M1 macrophages themselves are capable of production of IFN-γ as well.[28,29]

GRANULOCYTES

Granulocytes, as the agents of the innate immune system, which mediate inflammation, take part in the first steps of immune response against tumors. They are cytotoxic operators that act through release of substances such as inflammatory cytokines, ROS, cathepsin G, and azurocidin.[30] They may also play a role in cancer progression by induction of angiogenesis and metastasis.[31]

Neutrophils by using ROS and Fas/Fas ligand act as an antitumor immune cell. Eosinophils were observed in many cancer types specimens, while their anti- or protumoral activities are not obvious.[27] Basophils are the third type of granulocytes, which mainly have a role in allergic reactions. However, they have effects on antitumor immune response and act as APCs as well.[32]

Interestingly, granulocytes are among the significant effectors activated in responses to DNA vaccine. Moreover, there are reports of the associations between the dense infiltration of granulocytes and clinical responses to Bacillus Calmette-Guérin and autologous cancer cell vaccines secreting GM-CSF.[33–35]

B LYMPHOCYTES

B lymphocytes are the cells of adaptive immune system, which have either anti- or protumoral activities. They are capable of producing not only antibodies (Abs) targeting tumor antigens but also cytokines, which play a role in T cell functions. The antitumorigenic B cells have direct cytotoxic effect by release of granzyme B and indirect functions against cancer through antibody-dependent cell cytotoxicity and complement-dependent cytotoxicity, which both are mediated by antibodies secreted from B cells.[36]

They can also act as APCs in tumor microenvironment in case of failure of DCs in antigen presentation. The presence of B cells along with CD8+ T cells has been reported as a positive prognostic factor for survival of patients with ovarian cancer.[37]

T HELPER LYMPHOCYTES

T cells are the next cells of the immune system. They are divided into classical or αβ T cells and γδ T cells. The classical T cells have two main classes; i.e., CD4+ or T helper cells, often recognizing peptides presented by MHC-I, and CD8+ or cytotoxic T cells, which recognize peptides presented by MHC-II. MHC class I is located on the surface of all nucleated cells, and MHC class II is located only on APCs.[38]

According to the cytokines in tumor microenvironment, different types of CD4+ effector cells are originated from naïve CD4+ T cells. T helper (Th) cells are one of the routine kinds of this differentiation. Th1 cells are involved in cell-mediated immune responses developed by production of the cytokines such as IFN-γ, and Th2 cells are responsible for humoral or antibody dependent responses, which are developed by production of cytokines such as IL-4, IL-5, and IL-13.[39] Both Th1 and Th2 cells are involved in antitumor immunity. However, Th1 dominant immunity is vital for developing memory, which is required for providing a specific cytotoxic antitumor response.[40] IL-17+ T cells are CD4+ (Th17) or CD8+ (Tc17) T cells, which secrete IL-17. Both pro- and antitumor effects have been reported for IL-17+ cells.[27,41]

CYTOTOXIC T LYMPHOCYTES

CD8+ T cells are the cytotoxic agents of the adaptive immunity. They destroy tumor cells after they are activated by APCs. After termination of cytotoxic functions, programmed cell death is operated to hamper autoimmunity and damage to normal cells. Only a few CD8+ T cells (5%−10%) survive to continue as long-term memory T cells. They are categorized into central memory T cells (which are CD45RA−CCR7+) circulated in lymphoid tissues and effector memory T cells (which are CD45RA−CCR7−) circulated in peripheral tissues such as spleen.[42] Memory T cells show more rapid and robust immune responses against tumor cells than what do naïve T cells. Memory cells inhibit tumor cells from growth and metastasis; hence, induction of memory is an ambition of cancer immunotherapy.[21]

γδ T LYMPHOCYTES

The other type of T cells is γδ T cell, which contains a semi-invariant γδ TCR. They could destroy tumor cells from both hematological and solid tumors following recognition of stress ligands or tumor-derived phosphoantigens. They also produce TNF, and IFN-γ, which play an important role in tumor immunology.[27]

NATURAL KILLER T CELLS

NK T cells share antigens of both NK cell, CD161, and an invariant CD1d restricted TCR. These cells produce IFNγ and IL-4 to exert an indirect role in immunity against tumor. They also show cytotoxic effects by releasing cytotoxic agents, including perforin, granzyme B, and Fas ligand (FasL).[43]

NATURAL KILLER CELLS

NK cells are powerful cells of the innate immunity, which employ perforin and granzymes to destroy abnormal cells such as tumor cells. NK cells, which have activating and inhibitory receptors, recognize stress ligands on the tumor cells marked by lack of or decreased MHC molecules.[27,44]

INHIBITORY IMMUNE CELLS

TOLEROGENIC DCS

Based on lower expression of MHC and costimulatory molecules on the surface of immature DCs, tolerogenic DCs are unable to induce an immune response as powerful as mature DCs. They also are not capable of producing proinflammatory cytokines. These immature DCs form tolerogenic response either through removing antigen-specific T cells or through development of Tregs. They also secrete inhibitory cytokines such as IL-10, which hampers secretion of proinflammatory cytokines and debilitates NK cells and T cells, as well as TGF-β, which is necessary for induction and survival of Tregs.[45,46] Tolerogenic DCs might also be matured DCs, i.e., containing costimulatory molecules on their surface, but that can induce Tregs.[47]

ALTERNATIVELY ACTIVATED MACROPHAGES (M2S)

Alternatively activated macrophages (M2s) originate from monocytes developed under the effect of inhibitory cytokines such as IL-4, IL-13, IL-10, and TGF-β. M2s exert their inhibitory functions through angiogenesis, tumor invasion, and metastasis. They also attract Tregs to the tumor microenvironment through expressing CCL22, and can induce T cell apoptosis through expression of PD-L1 on their surface. There is positive feedback for induction of more M2s. IL-10, secreted by M2s, induces both M2 and Th2 development.[26]

MYELOID-DERIVED SUPPRESSOR CELLS

There are immature myeloid cells in tumor microenvironment, originate from either monocytes or granulocytes, which act as immunosuppressive agents. Coadministration of peripheral blood mononuclear cells with GM-CSF and IL-6, or GM-CSF and IL-1β, PGE2, TNF-α, and VEGF, results in induction of MDSCs, which use various mechanisms to hamper immune responses against tumors. They decrease nutrients vital for lymphocytes, interfere with IL-2 receptor signaling, interlope with lymphocyte trafficking, induce development of Tregs by CD40–CD40L linkage, and secrete IL-10 or TGF-β. They hamper T cell activation through arginase, inducible nitric oxide synthase (iNOS), ROS, or RNS.[20]

REGULATORY T CELLS

To prevent excessive functions of T cells and the following autoimmunity, Tregs exert their immunosuppressive activities. They restrict immune responses and lead to tolerance against self-antigens. Tregs have two main categories: (1) thymic-derived or tTregs, which are autoreactive T cells selected due to high avidity to interact with self-antigens in the thymus, and (2) peripheral or pTregs, originate from naïve CD4+ T cells, which do not present antigens optimally. Tregs classically contain the Forkhead Box P3 (FoxP3) transcription factor and the IL-2 receptor α chain (CD25), although, there are some exceptions with negative FoxP3. In patients with cancer, the number of Tregs is increased. They contain different inhibitory molecules to induce immunosuppression, for example, lymphocyte-activation gene 3, T-cell immunoglobulin and mucin-domain containing-3 (TIM-3/HAVCR2), glucocorticoid-induced tumor necrosis factor receptor, inducible T-cell costimulator, cytotoxic

T-lymphocyte associated protein 4 (CTLA-4), and programmed cell death protein 1. Tregs have indirect suppressive effects on effector T cells through decreasing local IL-2, the vital cytokine for the survival and proliferation of T cells. Production of immunosuppressive cytokines, such as IL-10 or TGF-β, is the other indirect way to induce suppression of T cells.[48,49]

REGULATORY B CELLS

Regulatory B cells (Bregs) play as an agent, which alters the balance of immune activating toward immunosuppressive status through secretion of IL-10 and TGF-β. Bregs showed effects on the function of T cells. Coculture of Bregs with stimulated autologous CD4+ T cells led to decrease in proliferation of CD4+ T cells. Suppression of Th1 differentiation is also abolished by CD40-stimulated human B cells and through IL-10 secretion. Tumor necrosis factor α is an immunosuppressive cytokine which recruits Bregs.[36]

IMMUNOEDITING HYPOTHESIS

It is not surprising to hear that one of the agents shaping the immunogenicity of tumors is the immune system. Evidence showed tumors in immunodeficient mice, which lack T, B, and NKT cells due to recombinase activating gene (RAG)-2 deficiency, are more immunogenic than tumors from immunocompetent models.[50] Understanding different steps of the tumor-immune system interference might lead to designing effective modalities to treat cancer.

ELIMINATION

Elimination is the first phase of the dynamic interaction between tumor and the immune system and it equally describes the immunosurveillance hypothesis. In this phase, the transformed cells must be recognized and destroyed by both arms of the immune system, i.e., the innate and adaptive immunities.

Studies on immunodeficient mice proved that the immune system is capable of suppressing tumor development, since the RAG-2 negative mice showed higher rate of tumor development and tumor immunogenicity compared to the wild-type mice.[50] Studies on this phase should provide answers to questions such as which central cells and effectors recognize and destroy the nascent tumor cells, and how the immune system recognizes the tumor cells.

RAG-2 is responsible for rearrangement of lymphocyte antigen receptors, and in case of lacking this gene, T, B, and NKT cells cannot be produced. Since this gene only functions in the lymphoid system and its lacking does not manipulate the DNA damage repair in other cells, RAG-2 negative mice are used as a potential model to evaluate lymphocyte-deficient mice and development of cancer.[51] After injection of carcinogen 3′-methylcholanthrene (MCA), a chemical substance used to induce sarcomas in mice, to RAG-$2^{-/-}$ model and wild-type control group, a higher rate of tumor development in lymphocyte-deficient mice was recorded.[52]

Experimental studies might show the innate immunity component to eliminate tumor; however, expansion of T cells is required and leads to preventing tumor development. Indeed, several players are involved in this phase including T cells, NK cells, cytotoxic agents such as IFN-γ, and perforin, and pathways like Fas/ FasL, and TNF-related apoptosis inducing ligand (TRAIL) and recognition molecules such as NKG2D.[53]

Recognition of tumor cells is the initiation of the elimination phase. It begins with making danger signals such as secretion of type I IFNs and involvement of other components secreted by or present on the surface of tumor cells such as DAMPs, RAG-2 in mice, and MHC class I chain-related protein A and B (MICA/B) in humans.[54–56]

Cells

More studies were carried out to determine which lymphocyte subtypes, influenced by RAG-2 deficiency, are responsible for antitumor response. Contribution of γδ and αβ T cells in the elimination phase was revealed by studies on TCR $\beta^{-/-}$ and TCR $\delta^{-/-}$ mice, which showed more susceptibility to tumor induction by MCA compared to wild-type mice.[57,58] NK cells and NKT cells are also the other players in the elimination phase. More susceptibility to develop tumors by MCA resulted in the models with blocked NK and NKT cells than wild-type mice.[59]

NK cells as a critical player of the innate immunity in the elimination phase directly destroy tumor cells by releasing lytic granules composed of perforin and granzymes. There are alternatives for the mentioned lytic components such as death receptor pathway through FasL and TRAIL, which must be activated by ligation with their relevant death receptors on the target cells.[44]

NK cells are also responsible for secretion of a huge amount of IFN-γ, which hampers tumor growth and shows antitumor activities through promoting the innate and adaptive immunities, stimulation of the tumoricidal capacities of macrophages, and Th1 polarization of CD4+ T cells. In addition, NK cells omit the immature DCs or support their maturation, leading to the efficient T cell priming by mature DCs.[60,61]

Interferons

Production of interferon gamma (IFN-γ) and capability to destroy tumor cells are the central functions of effector cells in the elimination phase. IFN-γ has direct and indirect antitumor functions. This pleiotropic cytokine induces tumor apoptosis, inhibits angiogenesis, and has a negative effect on tumor cell proliferation. It is a stronger agent for tumor immunogenicity induction compared to perforin.[62,63] IFN-γ also acts through activation of innate and adaptive immunity to destroy tumor cells.[62] As mentioned, NK cells are among the main sources of IFN-γ secretion, which is carried out under the influence of IL-12 and IL-18.[64]

IFN-γ, which is produced endogenously, establishes the cornerstone of the immunosurveillance phase.[65,66] In mice, which were deficient in IFN-γ or STAT1, a transcription factor necessary for the IFN-γ signaling, more susceptibility to tumor development induced by MCA was observed rather than in wild-type mice.[65] However, only deficiency in IL-12, IL-18, or TNF did not lead to spontaneous tumor development.[62]

IFN-γ takes part in increasing the expression of complexes and molecules playing a role in processing and presentation of tumor antigens such as MHC-I, transporters associated with TAP-1 and TAP-2, and the proteosomal components, low-molecular-weight proteins (LMP)-2 and LMP-7. IFN-γ causes induction of expression of multiple genes through activation of JAK–STAT pathways. Enhancement of tumor immunogenicity leads to tumor recognition by cytotoxic T lymphocytes (CTLs).[67]

To explore other cellular sources of this cytokine, mice with γδ T cells incapable of production of IFN-γ were compared with competent mice, and a greater incidence of tumor development in response to MCA or transplantation of B16 melanoma cells was observed in deficient mice.[58]

Type I IFNs including IFN-α and -β were also reported effective in protection against cancer in immunosurveillance phase. Administration of antibodies blocking IFN-α and -β increased tumor growth in wild-type mice with transplanted tumors.[68] Type I IFNs are among the "danger signals" alerting the immune system to presence of tumor cells. These cytokines activate DCs and launch cross-presentation of tumor antigens to T cells.[50]

Effects of type I IFNs on CD8α/CD103+ DCs leading to augmentation of cross-presentation is their main role to contribute in tumor rejection, and none of the other immune cells possessing type I IFNs showed contribution to this entity.[69]

Perforin

The other pivotal effector acting in the escape phase is perforin (pfp). Pfp performs tumoricidal activities through pore creation on the cell membrane after secretion from NK cells and CTLs.[70] Perforin-deficient mice showed two to three times higher tumor induction by MCA than did wild-type mice.[71] Aged perforin-deficient mice also developed spontaneous lymphomas eight times more than wild type mice.[72]

Either perforin or IFN-γ deletion increased the tumor burden in spontaneous metastasis models; however, the lack of both of these cytotoxic components reinforced the number of metastases in a similar level to mice with NK cell depletion. Therefore, IFN-γ and perforin-mediated cytotoxicity act separately but together perform equal to the antimetastatic effects of NK cells.[66]

Tumor Cell Recognition

Distinguishing the nascent primary tumor cells is the leading and substantial step for the whole immunoediting process. The tumor cells were originally normal cells of the body and were known as "self". Therefore, how are they identified among other normal cells? Tumor antigens are the answer. There are some molecules, mainly peptides or proteins, on the surface of the tumor cells, which are signs for immune recognition. There are five main categories of tumor antigens: (1) differentiation antigens expressed by limited normal cells and tumor cells, like, melanocyte differentiation antigens, Melan-A/MART-1, gp-100, NY-BR-1, and CEA; (2) overexpressed or amplified antigens, such as PSA and HER-2/neu; (3) mutational antigens, such as oncogenic forms of p53 and KRAS; (4) cancer-testis (CT) antigens, e.g., MAGE and NY-ESO-1; and (5) viral antigens, e.g., HPV. Among the mentioned categories, the last three are tumor specific, as they are not expressed by the normal cells. CT antigens are an exception expressed by cancerous, germline, and trophoblastic cells.[73,74]

Moreover, some molecules function as signs for tumor recognition such as NKG2D-activating receptor, which is detected on NK cells, $\gamma\delta$ T cells, and CD8$\alpha\beta$ T cells.[75] Hyperploidy is the other state leading to immunogenicity and elimination of the tumor, and especially tetraploidy, leading to oncogenesis, results in tumor immunogenicity.[76]

A comparison between immunogenicity of the tumor grown in immunocompetent or immunodeficient mice demonstrated that immunodeficiency leads to development of the tumors, which are more immunogenic. Transplantation of these two kinds of tumors resulted in tumor rejection in 40% of tumors developed in the immunodeficient mice while none of the tumors grown in the immunocompetent mice were rejected by the wild-type mice.[52]

If the tumor is eliminated, the immunoediting is over; however, it is not the final phase in case that the immune system fails to immunoedit the tumor.

Immunotherapy could also alter the immunogenicity and immunoediting of the tumor. Recently it has been reported that in a patient with melanoma whose tumor at first was positive for the antigens NY-ESO-1, MAGE-C1 and Melan-A, administration of a vaccine against NY-ESO-1 led to the progression of tumors that were deficient in NY-ESO-1 but not in MHC Class I, and the other tumor antigens, i.e., MAGE-C1 and Melan-A.[77] In another report of the phase I clinical trial in patients with melanoma assessed by an NY-ESO vaccine, analysis of the tumors in relapsed patients demonstrated loss of NY-ESO or MHC class I.[78]

EQUILIBRIUM

If the elimination phase fails to eradicate the tumor completely, the next phase is equilibrium, in which more genetic and epigenetic alterations exist and resistant tumors develop. Indeed, it is a functional dormancy status between the elimination and escape phases that controls the tumor but cannot eradicate it. This phase prevents appearance of clinical manifestations of cancer.[79,80]

The molecular interactions underlying the equilibrium phase have been less discovered because it is not simple to model in mice. Albeit, there have been some attempts comparing the equilibrium with elimination or escape phases to understand what happens during the equilibrium and how it is different from the two other phases and how it could shift to them. In a study using low doses of MCA inducing occult tumor, the equilibrium phase has been shown mechanistically distinct from the two other phases. It has been reported that the adaptive Th1-like immunity plays a role in immune-mediated dormancy of sarcoma.[81] T helper-1 cells by production of IFN-γ and TNF could also cease the outgrowth of tumor by induction of senescence in tumor cells, which is a cytokine-dependent senescence and not an oncogene-dependent senescence.[82]

The equilibrium phase has been reported to be a long-lasting period, in which there have been antitumor effects for IL-12 and protumor activities for IL-23 and IL-10. The other cytokines involved in the elimination phase of the MCA induced cancer, such as IL-4, IL-17, TNF, and IFN$\alpha\beta$, did not show accountability for the equilibrium phase.[83] Comparing the cellular components of the equilibrium and escape phases showed a balance between the immune effector and immunosuppressive states in the equilibrium and an immunosuppressive dominancy during the escape. High proportions of effector or memory CTLs, NK cells, and $\gamma\delta$T cell and low proportion of mononuclear myeloid-derived suppressor cells and Tregs contribute to the relative balance. Two cell count ratios are associated with maintaining tumor dormancy as well, the high CD8+/CD4+ and CD8+/Treg cell ratios.[84]

Early diagnosis of occult cancer in equilibrium expectantly increases the advantages of the cancer immunotherapy. Development of modern technologies such as imaging, genetics, and biomarker tests might change the current status of cancer diagnosis in occult states.

ESCAPE

Tumor escape happens when the immune system is not capable enough to eradicate tumors, which carry accumulated genetic and epigenetic alterations and use several mechanisms to be the victorious of the immunoediting process. Following the tumor outgrowth, clinical manifestations of cancer develop.

The mechanisms interfering in shaping robust antitumor immune responses are mainly divided into the following categories: (1) defective tumor antigen processing or presentation, (2) lack of activating mechanisms, (3) inhibitory mechanisms and immunosuppressive state, and (4) resistant tumor cells. Herein, we will discuss each category separately.

Defects in Tumor Antigen Processing, Presentation, and Recognition

As mentioned before, some categories of tumor antigens are shared with the antigens of normal cells, which have "self" nature and are recognized by low-affinity T cells against self-antigens. If the antigen/MHC-I complexes are abandoned, the condition induced through vaccines, the low-affinity T cells become capable of eliminating tumor cells.[85]

The tumor antigens might be present in a new form due to the genetic instability, mutation of the tumor and escape from immune system. Epitope-negative tumor cells remain hidden and consequently resistant to the immune rejection. They have been developed following the elimination of epitope-positive tumor cells,[86] similar to Darwin's theory of natural selection. Loss of MHC-I during progression and metastasis of the tumor contributes to immune escape as well. Similarly, T cells give rise to the number of MHC-I–negative tumors by eradicating MHC-I–positive tumors.[87] Although immunotherapy leads to the upregulation of MHC-I,[88] and IFN-γ increases the expression of MHC-I as well, this does not necessarily lead to tumor rejection.[89] Antigen processing machinery in DC is the other factor that might be decreased due to the tumor progression.[90] However, it is reversible by incubation with cytokines such as TNF-α, IFN-γ, IL-6, IL1β, and IFN-α.[91]

Lack of Activating Mechanisms

In addition to the main signal transduced through presentation of peptide/MHC-I complex to the TCRs, costimulatory signals are required to recognize the tumor antigen. In the case of lacking costimulatory molecules such as CD28 on the surface T cells and CD80 on the surface of APCs, T cell activation does not occur.[92,93]

Tumor cell–induced apoptosis of T cells results in decreased immune responses against cancer. It is executed by interaction of Fas and FasL on the surface of T cells and tumor cells, respectively.[94] Tumor cells by overexpression of gangliosides also inhibit T cell proliferation and cytokine secretion.[95]

Inhibitory Mechanisms and Immunosuppressive State

When the costimulatory signals are replaced by inhibitory signals, T cell activation fails. CTLA-4 is an inhibitory molecule on the surface of T cells, which competes with CD28 to bind the B7 proteins, CD80 and CD86. There is also evidence of development of inhibitory mechanisms by CTLA-4 in the absence of CD28.[96] In this case, CTLA-4 transduces a negative signaling independent of CD28, resulting in rapid T cell inhibition. However, the competitive function of CTLA-4 with the CD28 costimulatory function leads to T cell anergy.[97]

There are some inhibitory immune cells, such as Tregs and MDSC, that can alter the outcomes of an immune response against tumors. As mentioned previously, Tregs are CD4+ CD25 high FOXP3+ T cells, which are increased in tumors and the peripheral circulation of patients with cancer, probably due to the self-nature of tumor antigens and to induce tolerance against them. Tregs perform their immunosuppressive activities through different approaches such as production of inhibitory cytokines like IL-10 and TGF-β1,[98] enzymatic degradation of ATP to immunosuppressive adenosine,[99] modulation of apoptosis of T cells by Fas/FasL pathway,[100] and production of granzyme/perforin.[101]

There is an increase in the number of the other immunosuppressive cell, MDSCs, in patients with cancer. Some factors such as IFNγ, TGFβ, ligands for TLRs, IL-4, and IL-13, which are secreted by the activated T cells and the tumor stroma, actuate the MDSCs.[102] Their immunosuppressive activities are accomplished by overexpression of immune inhibitory factors such as arginase and iNOS. Increased production of ROS by the MDSCs is the other mechanism to suppress immunity against tumor.[103] IDO, expressed by MDSCs, also functions to hamper starting Ag-specific immune response and cytotoxicity effects of T cells via decrease in local tryptophan, an essential amino acid for T cell differentiation. It gives rise to the frequency of Tregs as well.[104,105]

The tumor-associated chronic inflammation makes the tumor microenvironment immunosuppressive by polarization of immune cells toward a phenotype favoring tumor outgrowth like development of tumor-associated macrophages and MDSCs, which are myeloid cells that remain immature due to the inflammation. Production of cytokines such as TNF-α, which shows carcinogenesis via the nuclear factor−kappaB pathway, IFN-γ, IL-6, IL-17, IL-22, and IL-23, shape the immunosuppressive microenvironment suitable for tumor progression.[106−109]

Resistant Tumor Cells

As the pathologic process of cancer progresses, more genomic instabilities are accumulated in the tumor cells, which make them more resistant to immune responses and immunotherapy. Cancer stem cells are the other cause of resistance to immunity and cancer treatment, since they are multipotent cells, well-preserved from the immune system, with self-renewal capability, and live for a long time.[110]

REFERENCES

1. Medzhitov R. Recognition of microorganisms and activation of the immune response. *Nature* 2007; **449**(7164):819−26.
2. Newton K, Dixit VM. Signaling in innate immunity and inflammation. *Cold Spring Harb Perspect Biol* 2012;**4**(3):a006049.
3. Kawai T, Akira S. The role of pattern-recognition receptors in innate immunity: update on Toll-like receptors. *Nat Immunol* 2010;**11**(5):373−84.
4. Vesely MD, et al. Natural innate and adaptive immunity to cancer. *Annu Rev Immunol* 2011;**29**:235−71.
5. Janeway Jr CA, T P, Walport M, et al. *Immunobiology: the immune system in health and disease*. 5th ed. , New York: Garland Science; 2001.
6. Galluzzi L, et al. Immunogenic cell death in cancer and infectious disease. *Nat Rev Immunol* 2017;**17**(2): 97−111.
7. Kershaw MH, Westwood JA, Darcy PK. Gene-engineered T cells for cancer therapy. *Nat Rev Cancer* 2013; **13**(8):525−41.
8. Ossendorp F, et al. Specific T helper cell requirement for optimal induction of cytotoxic T lymphocytes against major histocompatibility complex class II negative tumors. *J Exp Med* 1998;**187**(5):693−702.
9. Janssen EM, et al. CD4+ T cells are required for secondary expansion and memory in CD8+ T lymphocytes. *Nature* 2003;**421**(6925):852−6.
10. Melief CJM, et al. Therapeutic cancer vaccines. *J Clin Invest* 2015;**125**(9):3401−12.
11. Munn DH, Mellor AL. Indoleamine 2,3 dioxygenase and metabolic control of immune responses. *Trends Immunol* 2013;**34**(3):137−43.
12. Lippitz BE. Cytokine patterns in patients with cancer: a systematic review. *Lancet Oncol* 2013;**14**(6): e218−28.

13. Gorbachev AV, Fairchild RL. Regulation of chemokine expression in the tumor microenvironment. *Crit Rev Immunol* 2014;**34**(2):103–20.
14. Nagarsheth N, Wicha MS, Zou W. Chemokines in the cancer microenvironment and their relevance in cancer immunotherapy. *Nat Rev Immunol* 2017;**17**(9):559–72.
15. Siegal FP, et al. The nature of the principal type 1 interferon-producing cells in human blood. *Science* 1999; **284**(5421):1835–7.
16. Palucka K, Ueno H, Banchereau J. Recent developments in cancer vaccines. *J Immunol* 2011;**186**(3): 1325–31.
17. Segura E, Durand M, Amigorena S. Similar antigen cross-presentation capacity and phagocytic functions in all freshly isolated human lymphoid organ-resident dendritic cells. *J Exp Med* 2013;**210**(5):1035–47.
18. Collin M, McGovern N, Haniffa M. Human dendritic cell subsets. *Immunology* 2013;**140**(1):22–30.
19. Klechevsky E, et al. Functional specializations of human epidermal Langerhans cells and CD14+ dermal dendritic cells. *Immunity* 2008;**29**(3):497–510.
20. Guo C, et al. Therapeutic cancer vaccines: past, present, and future. *Adv Cancer Res* 2013;**119**:421–75.
21. Raval RR, et al. Tumor immunology and cancer immunotherapy: summary of the 2013 SITC primer. *J. Immunother. Cancer* 2014;**2**:14.
22. Jin Y, et al. Regulation of anti-inflammatory cytokines IL-10 and TGF-β in mouse dendritic cells through treatment with Clonorchis sinensis crude antigen. *Exp Mol Med* 2014;**46**(1):e74.
23. Wang H, et al. Pro-tumor activities of macrophages in the progression of melanoma. *Hum Vaccines Immunother* 2017;**13**(7):1556–62.
24. Sica A, et al. Macrophage polarization in tumour progression. *Semin Cancer Biol* 2008;**18**(5):349–55.
25. Teicher BA, Fricker SP. CXCL12 (SDF-1)/CXCR4 pathway in cancer. *Clin Cancer Res* 2010;**16**(11): 2927–31.
26. Quatromoni JG, Eruslanov E. Tumor-associated macrophages: function, phenotype, and link to prognosis in human lung cancer. *Am J Transl Res* 2012;**4**(4):376–89.
27. Darcy PK, et al. Manipulating immune cells for adoptive immunotherapy of cancer. *Curr Opin Immunol* 2014;**27**:46–52.
28. Schindler H, et al. The production of IFN-γ by IL-12/IL-18-activated macrophages requires STAT4 signaling and is inhibited by IL-4. *J Immunol* 2001;**166**(5):3075–82.
29. Zwirner NW, Ziblat A. Regulation of NK cell activation and effector functions by the IL-12 family of cytokines: the case of IL-27. *Front Immunol* 2017;**8**:25.
30. Jinushi M, Dranoff G. Immunosurveillance: innate and adaptive antitumor immunity A2. In: Prendergast GC, Jaffee EM, editors. *Cancer immunotherapy.* Burlington: Academic Press; 2007. p. 29–41 (Chapter 3).
31. Gregory AD, McGarry Houghton A. Tumor-associated neutrophils: new targets for cancer therapy. *Canc Res* 2011;**71**(7):2411–6.
32. Schneider E, et al. Basophils: new players in the cytokine network. *Eur Cytokine Netw* 2010;**21**(3):142–53.
33. Curcio C, et al. Nonredundant roles of antibody, cytokines, and perforin in the eradication of established Her-2/neu carcinomas. *J Clin Invest* 2003;**111**(8):1161–70.
34. Soiffer R, et al. Vaccination with irradiated, autologous melanoma cells engineered to secrete granulocyte-macrophage colony-stimulating factor by adenoviral-mediated gene transfer augments antitumor immunity in patients with metastatic melanoma. *J Clin Oncol* 2003;**21**(17):3343–50.
35. Suttmann H, et al. Neutrophil granulocytes are required for effective Bacillus Calmette-Guerin immunotherapy of bladder cancer and orchestrate local immune responses. *Cancer Res* 2006;**66**(16):8250–7.
36. Fremd C, et al. B cell-regulated immune responses in tumor models and cancer patients. *OncoImmunology* 2013;**2**(7):e25443.

37. Nielsen JS, et al. $CD20^{+}$ tumor-infiltrating lymphocytes have an atypical $CD27^{-}$ memory phenotype and together with $CD8^{+}$ T cells promote favorable prognosis in ovarian cancer. *Clin Cancer Res* 2012;**18**(12):3281–92.
38. Ashton-Rickardt PG, Tonegawa S. A differential-avidity model for T-cell selection. *Immunol Today* 1994;**15**(8):362–6.
39. Seledtsov VI, Seledtsova GV. A possible role of pre-existing IgM/IgG antibodies in determining immune response type. *Immunol Cell Biol* 1997;**75**(2):176–80.
40. Nishimura T, et al. The critical role of Th1-dominant immunity in tumor immunology. *Cancer Chemother Pharmacol* 2000;**46**(Suppl.):S52–61.
41. Srenathan U, Steel K, Taams LS. IL-17+ CD8+ T cells: differentiation, phenotype and role in inflammatory disease. *Immunol Lett* 2016;**178**:20–6.
42. Obhrai JS, et al. Effector T cell differentiation and memory T cell maintenance outside secondary lymphoid organs. *J Immunol* 2006;**176**(7):4051–8.
43. Gansuvd B, et al. Human umbilical cord blood NK T cells kill tumors by multiple cytotoxic mechanisms. *Hum Immunol* 2002;**63**(3):164–75.
44. Screpanti V, et al. A central role for death receptor-mediated apoptosis in the rejection of tumors by NK cells. *J Immunol* 2001;**167**(4):2068–73.
45. Manicassamy S, Pulendran B. Dendritic cell control of tolerogenic responses. *Immunol Rev* 2011;**241**(1):206–27.
46. Yoo S, Ha S-J. Generation of tolerogenic dendritic cells and their therapeutic applications. *Immune Network* 2016;**16**(1):52–60.
47. Akbari O, DeKruyff RH, Umetsu DT. Pulmonary dendritic cells producing IL-10 mediate tolerance induced by respiratory exposure to antigen. *Nat Immunol* 2001;**2**(8):725–31.
48. Chaudhary B, Elkord E. Regulatory T cells in the tumor microenvironment and cancer progression: role and therapeutic targeting. *Vaccines* 2016;**4**(3):28.
49. Li Z, et al. FOXP3+ regulatory T cells and their functional regulation. *Cell Mol Immunol* 2015;**12**:558.
50. Schreiber RD, Old LJ, Smyth MJ. Cancer immunoediting: integrating immunity's roles in cancer suppression and promotion. *Science* 2011;**331**(6024):1565–70.
51. Shinkai Y, et al. RAG-2-deficient mice lack mature lymphocytes owing to inability to initiate V(D)J rearrangement. *Cell* 1992;**68**(5):855–67.
52. Shankaran V, et al. IFNgamma and lymphocytes prevent primary tumour development and shape tumour immunogenicity. *Nature* 2001;**410**(6832):1107–11.
53. Mittal D, et al. New insights into cancer immunoediting and its three component phases—elimination, equilibrium and escape. *Curr Opin Immunol* 2014;**27**:16–25.
54. Matzinger P. Tolerance, danger, and the extended family. *Annu Rev Immunol* 1994;**12**:991–1045.
55. Sims GP, et al. HMGB1 and RAGE in inflammation and cancer. *Annu Rev Immunol* 2010;**28**:367–88.
56. Guerra N, et al. NKG2D-deficient mice are defective in tumor surveillance in models of spontaneous malignancy. *Immunity* 2008;**28**(4):571–80.
57. Girardi M, et al. Regulation of cutaneous malignancy by gammadelta T cells. *Science* 2001;**294**(5542):605–9.
58. Gao Y, et al. Gamma delta T cells provide an early source of interferon gamma in tumor immunity. *J Exp Med* 2003;**198**(3):433–42.
59. Smyth MJ, Crowe NY, Godfrey DI. NK cells and NKT cells collaborate in host protection from methylcholanthrene-induced fibrosarcoma. *Int Immunol* 2001;**13**(4):459–63.
60. Martin-Fontecha A, et al. Induced recruitment of NK cells to lymph nodes provides IFN-gamma for T(H)1 priming. *Nat Immunol* 2004;**5**(12):1260–5.

61. Celada A, et al. Evidence for a gamma-interferon receptor that regulates macrophage tumoricidal activity. *J Exp Med* 1984;**160**(1):55–74.
62. Street SEA, et al. Suppression of lymphoma and epithelial malignancies effected by interferon γ. *J Exp Med* 2002;**196**(1):129–34.
63. Dighe AS, et al. Enhanced in vivo growth and resistance to rejection of tumor cells expressing dominant negative IFN gamma receptors. *Immunity* 1994;**1**(6):447–56.
64. Guillerey C, Smyth MJ. NK cells and cancer immunoediting. *Curr Top Microbiol Immunol* 2016;**395**: 115–45.
65. Kaplan DH, et al. Demonstration of an interferon gamma-dependent tumor surveillance system in immunocompetent mice. *Proc Natl Acad Sci U S A* 1998;**95**(13):7556–61.
66. Street SE, Cretney E, Smyth MJ. Perforin and interferon-gamma activities independently control tumor initiation, growth, and metastasis. *Blood* 2001;**97**(1):192–7.
67. Ritter M, et al. Lytic susceptibility of target cells to cytotoxic T cells is determined by their constitutive major histocompatibility complex class I antigen expression and cytokine-induced activation status. *Immunology* 1994;**81**(4):569–77.
68. Gresser I, Belardelli F. Endogenous type I interferons as a defense against tumors. *Cytokine Growth Factor Rev* 2002;**13**(2):111–8.
69. Diamond MS, et al. Type I interferon is selectively required by dendritic cells for immune rejection of tumors. *J Exp Med* 2011;**208**(10):1989–2003.
70. Osińska I, Popko K, Demkow U. Perforin: an important player in immune response. *Cent-Eur J Immunol* 2014;**39**(1):109–15.
71. van den Broek ME, et al. Decreased tumor surveillance in perforin-deficient mice. *J Exp Med* 1996;**184**(5): 1781–90.
72. Smyth MJ, et al. Perforin-mediated cytotoxicity is critical for surveillance of spontaneous lymphoma. *J Exp Med* 2000;**192**(5):755–60.
73. Old LJ. Cancer vaccines 2003: opening address. *Cancer Immun Arch* 2003;**3**(Suppl. 2).
74. Vigneron N. Human tumor antigens and cancer immunotherapy. *BioMed Res Int* 2015;**2015**:17.
75. Bauer S, et al. Activation of NK cells and T cells by NKG2D, a receptor for stress-inducible MICA. *Science* 1999;**285**(5428):727–9.
76. Senovilla L, et al. An immunosurveillance mechanism controls cancer cell ploidy. *Science* 2012;**337**(6102): 1678–84.
77. von Boehmer L, et al. NY-ESO-1-specific immunological pressure and escape in a patient with metastatic melanoma. *Cancer Immun* 2013;**13**:12.
78. Nicholaou T, et al. Immunoediting and persistence of antigen-specific immunity in patients who have previously been vaccinated with NY-ESO-1 protein formulated in ISCOMATRIX. *Cancer Immunol Immunother* 2011;**60**(11):1625–37.
79. Wheelock EF, Weinhold KJ, Levich J. The tumor dormant state. *Adv Cancer Res* 1981;**34**:107–40.
80. Uhr JW, et al. Cancer dormancy: studies of the murine BCL1 lymphoma. *Cancer Res* 1991;**51**(18 Suppl.): 5045s–53s.
81. Koebel CM, et al. Adaptive immunity maintains occult cancer in an equilibrium state. *Nature* 2007; **450**(7171):903–7.
82. Braumuller H, et al. T-helper-1-cell cytokines drive cancer into senescence. *Nature* 2013;**494**(7437):361–5.
83. Teng MWL, et al. Opposing roles for IL-23 and IL-12 in maintaining occult cancer in an equilibrium state. *Canc Res* 2012;**72**(16):3987–96.
84. Wu X, et al. Immune microenvironment profiles of tumor immune equilibrium and immune escape states of mouse sarcoma. *Cancer Lett* 2013;**340**(1):124–33.

85. Houghton AN, Guevara-Patiño JA. Immune recognition of self in immunity against cancer. *J Clin Invest* 2004;**114**(4):468–71.
86. Khong HT, Wang QJ, Rosenberg SA. Identification of multiple antigens recognized by tumor-infiltrating lymphocytes from a single patient: tumor escape by antigen loss and loss of MHC expression. *J Immunother* 2004;**27**(3):184–90.
87. Algarra I, Cabrera T, Garrido F. The HLA crossroad in tumor immunology. *Hum Immunol* 2000;**61**(1): 65–73.
88. Carretero R, et al. Regression of melanoma metastases after immunotherapy is associated with activation of antigen presentation and interferon-mediated rejection genes. *Int J Canc* 2012;**131**(2):387–95.
89. Hallermalm K, et al. Modulation of the tumor cell phenotype by IFN-γ results in resistance of uveal melanoma cells to granule-mediated lysis by cytotoxic lymphocytes. *J Immunol* 2008;**180**(6):3766–74.
90. Seliger B, Maeurer MJ, Ferrone S. Antigen-processing machinery breakdown and tumor growth. *Immunol Today* 2000;**21**(9):455–64.
91. Whiteside TL, et al. Antigen-processing machinery in human dendritic cells: up-regulation by maturation and down-regulation by tumor cells. *J Immunol* 2004;**173**(3):1526–34.
92. Foreman KE, et al. Expression of costimulatory molecules CD80 and/or CD86 by a Kaposi's sarcoma tumor cell line induces differential T-cell activation and proliferation. *Clin Immunol* 1999;**91**(3):345–53.
93. Torres L, et al. Loss of the CD28 costimulatory molecules on the immune subsets of TCD4+ cells in prostate cancer elderly patients. *J Clin Oncol* 2016;**34**(15 Suppl.):e16612.
94. Gastman BR, et al. Tumor-induced apoptosis of T lymphocytes: elucidation of intracellular apoptotic events. *Blood* 2000;**95**(6):2015–23.
95. Frey AB. Cancer-induced signaling defects in antitumor T cells. *Immunol Rev* 2008;**222**:192–205.
96. McCoy KD, Le Gros G. The role of CTLA-4 in the regulation of T cell immune responses. *Immunol Cell Biol* 1999;**77**:1.
97. Carreno BM, et al. CTLA-4 (CD152) can inhibit T cell activation by two different mechanisms depending on its level of cell surface expression. *J Immunol* 2000;**165**(3):1352–6.
98. Strauss L, et al. A unique subset of CD4+CD25highFoxp3+ T cells secreting interleukin-10 and transforming growth factor-beta1 mediates suppression in the tumor microenvironment. *Clin Cancer Res* 2007; **13**(15 Pt 1):4345–54.
99. Schuler PJ, et al. Human CD4+ CD39+ regulatory T cells produce adenosine upon co-expression of surface CD73 or contact with CD73+ exosomes or CD73+ cells. *Clin Exp Immunol* 2014;**177**(2):531–43.
100. Strauss L, Bergmann C, Whiteside TL. Human circulating CD4(+)CD25(high)Foxp3(+)regulatory T cells kill autologous CD8(+)but not CD4(+)Responder cells by Fas-mediated apoptosis. *J Immunol* 2009; **182**(3):1469–80.
101. Cao X, et al. Granzyme B and perforin are important for regulatory T cell-mediated suppression of tumor clearance. *Immunity* 2007;**27**(4):635–46.
102. Gabrilovich DI, Nagaraj S. Myeloid-derived-suppressor cells as regulators of the immune system. *Nat Rev Immunol* 2009;**9**(3):162–74.
103. Kusmartsev S, et al. Antigen-specific inhibition of CD8+ T cell response by immature myeloid cells in cancer is mediated by reactive oxygen species. *J Immunol* 2004;**172**(2):989–99.
104. Lee GK, et al. Tryptophan deprivation sensitizes activated T cells to apoptosis prior to cell division. *Immunology* 2002;**107**(4):452–60.
105. Curti A, et al. Modulation of tryptophan catabolism by human leukemic cells results in the conversion of $CD25^-$ into $CD25^+$ T regulatory cells. *Blood* 2007;**109**(7):2871–7.
106. Wang D, DuBois RN. Immunosuppression associated with chronic inflammation in the tumor microenvironment. *Carcinogenesis* 2015;**36**(10):1085–93.

107. Pikarsky E, et al. NF-kappaB functions as a tumour promoter in inflammation-associated cancer. *Nature* 2004;**431**(7007):461–6.
108. Noy R, Pollard JW. Tumor-associated macrophages: from mechanisms to therapy. *Immunity* 2014;**41**(1): 49–61.
109. Umansky V, et al. Myeloid-derived suppressor cells and tumor escape from immune surveillance. *Semin Immunopathol* 2017;**39**(3):295–305.
110. Bruttel VS, Wischhusen J. Cancer stem cell immunology: key to understanding tumorigenesis and tumor immune escape? *Front Immunol* 2014;**5**:360.

CHAPTER

IMMUNOTHERAPEUTIC APPROACHES IN CANCER

2

Mahsa Keshavarz-Fathi[1,2,3], **Nima Rezaei**[3,4,5]

School of Medicine, Tehran University of Medical Sciences, Tehran, Iran[1]; *Cancer Immunology Project (CIP), Universal Scientific Education and Research Network (USERN), Tehran, Iran*[2]; *Research Center for Immunodeficiencies, Children's Medical Center, Tehran University of Medical Sciences, Tehran, Iran*[3]; *Department of Immunology, School of Medicine, Tehran University of Medical Sciences, Tehran, Iran*[4]; *Network of Immunity in Infection, Malignancy and Autoimmunity (NIIMA), Universal Scientific Education and Research Network (USERN), Tehran, Iran*[5]

HISTORY

In 1891, Coley examined a bacterial toxin consisting of streptococcal organisms for treatment of a patient with sarcoma and observed shrinkage of the tumor. Although he was not the first person who stated the relationship between infection and tumor regression, he carried out the first systematic study evaluating cancer immunotherapy and recorded evidence of such effect. Indeed, Coley used the toxin to cause erysipelas and subsequently stimulate the immune system. Coley's toxin was the first example of a cancer vaccine, which showed positive results in some cases and side effects as well. He could be named as the "father of immunotherapy," who started the long path of this concept, i.e. treating cancer by evoking the immune system.[1,2]

Later on, understanding of the immune functions such as Paul Ehrlich's explanation of the role of immune system in recognition and suppression of tumor led to progress in studying the interaction between the immune system and cancer.[3] Developing immunosurveillance and immunoediting hypothesis were the next stages for configuration of cancer immunotherapy.

Immunosurveillance is a state of tumor control by the immune system, which was first proposed by Macfarlane Burnet and Lewis Thomas. Burnet, regarding the immune tolerance mechanisms, recognized that cancerous neo-antigens ultimately protected the body against neoplasm. Thomas, regarding the rejected homografts, believed that there might be similar mechanisms to protect the body against neoplasm.[4] Afterwards, it was proven that immunosurveillance is not the whole story of the complex interaction of the immune system and neoplasm. This led to proposing the immunoediting hypothesis, which includes three phases: elimination (immunosurveillance), immune equilibrium, and immune scape. The hypothesis describes different statuses, in which either the tumor or the immune system overcomes each other. The new findings advanced and expanded the use of cancer immunotherapy. Until now, many approaches have been developed to treat cancer by targeting the immune system and its interaction with tumors. Here we will review various types of cancer immunotherapies and discuss interference of the mechanisms involved in the interaction between immune system and tumor cells to cease the growth and invasion of cancer.

Vaccines for Cancer Immunotherapy. https://doi.org/10.1016/B978-0-12-814039-0.00002-3

APPROACHES OF CANCER IMMUNOTHERAPY

Understanding the interactions between the tumor and the immune system has paved the way of translational medicine and designing various immunotherapeutic approaches for cancer treatment, which applies mechanisms distinct from conventional therapies such as surgery, chemotherapy, and radiotherapy. Depending on the type of immunotherapy, different efficacies and toxicities have been yielded.

Immunotherapeutic approaches to treat cancer can be classified from different views. One of the old classifications is done based on active or passive approaches alone or integrated with specific and nonspecific approaches according to the specificity to the tumor antigen and obviously the tumor type. Targeting inhibitory mechanisms or empowering the activating mechanisms provides the basis for the next type of classification, which can be explained according to the various stages of tumor/immune system interaction. Here we will discuss the rationale for a variety of immunotherapies against cancer and review the U.S. Food and Drug Administration (FDA)-approved immunotherapeutics for cancer in Table 2.1.

Table 2.1 Immunotherapies Approved by FDA for Cancer

Immunotherapy Drug	Category Mechanism of Action	Indication(s)	Year of Approval
Axicabtagene ciloleucel (Yescarta)	CAR T cell therapy Targeting CD19 and providing first and costimulatory signaling through CD3ζ and CD28	DLBCL: in patients who had follicular lymphoma, primary mediastinal large B-cell lymphoma, and high-grade B-cell lymphoma[5]	2017
Inotuzumab ozogamicin (Besponsa)	Antibody–drug conjugate targeting CD22 to deliver a toxic agent	B-cell ALL: for adults with Philadelphia chromosome, progressed despite targeted therapy[6]	2017
Tisagenlecleucel (Kymriah)	CAR T-cell therapy Targeting CD19 and providing first and costimulatory signaling through CD3ζ and 4-1BB	Adult patients with relapsed or refractory large B-cell lymphoma after two or more lines of systemic therapy including DLBCL not otherwise specified, high grade B-cell lymphoma and DLBCL arising from follicular lymphoma (NCT02445248)	2018
		B-cell ALL: not responding to treatment or relapsed ≥2 times, and up to age of 25 (NCT02435849)	2017
Durvalumab (Imfinzi)	Immune checkpoint inhibitor Blockade of PD-L1	Unresectable stage III NSCLC not progressed following concurrent platinum-based chemotherapy and radiation therapy[7]	2018
		Locally advanced or metastatic bladder cancer progressed during or after platinum-based chemotherapy or within 12 months of neoadjuvant or adjuvant chemotherapy[8]	2017

Table 2.1 Immunotherapies Approved by FDA for Cancer—cont'd

Immunotherapy Drug	Category Mechanism of Action	Indication(s)	Year of Approval
Avelumab (Bavencio)	Immune checkpoint inhibitor Blockade of PD-L1	Locally advanced or metastatic bladder cancer progressed during or after platinum-based chemotherapy or within 12 months of neoadjuvant or adjuvant chemotherapy[9]	2017
Avelumab (Bavencio)	Immune checkpoint inhibitor Blockade of PD-L1	Metastatic Merkel cell carcinoma in patients older 3 years[10]	2017
Olaratumab (Lartruvo)	mAb Inhibitor of PDGFRα	Combined with doxorubicin for soft tissue sarcoma not curable by surgery or radiation therapy, and usually responding to anthracycline-based chemotherapy[11]	2016
atezolizumab (Tecentriq)	Immune checkpoint inhibitor Blockade of PD-L1	Locally advanced or metastatic urothelial carcinoma not eligible to for cisplatin-based chemotherapy (NCT02951767)	2017
		Metastatic NSCLC progressed during or after a first-line platinum-based chemotherapy without EGFR or ALK genomic tumor aberrations[12]	2016
		Locally advanced or metastatic urothelial carcinoma progressed during or after treatment with platinum-based chemotherapy[13]	2016
Elotuzumab (Empliciti)	mAb Targeting SLAMF7	Combined with lenalidomide and dexamethasone for multiple myeloma received 1–3 prior treatments[14]	2015
Necitumumab (Portrazza)	mAb Targeting EGFR	Combined with standard chemotherapy, for initial treatment of metastatic squamous NSCLC[15]	2015
Daratumumab (Darzalex)	mAb Targeting CD38	Combined with one of two other standard treatments in multiple myeloma progressed after only a single prior treatment[16,17]	2016
		Multiple myeloma received $\geq$3 prior treatments[18]	2015
Talimogene laherparepvec (T-VEC, or Imlygic)	Oncolytic virus therapy Selectively replicates in tumor cells and secretes GM-CSF	Recurrent melanoma lesions metastasized to the skin and lymph nodes[19]	2015
Dinutuximab (Unituxin)	mAb Targeting GD2	Combined with GM-CSF, IL-2, and 13-cis-retinoic acid, for high-risk neuroblastoma, with at least a partial response to first-line multiagent and multimodality therapy[20]	2015

Continued

Table 2.1 Immunotherapies Approved by FDA for Cancer—cont'd

Immunotherapy Drug	Category Mechanism of Action	Indication(s)	Year of Approval
Nivolumab (Opdivo)	Immune checkpoint inhibitor Blockade of PD-1	Advanced liver cancer (hepatocellular carcinoma) previously treated by sorafenib (Nexavar) (NCT01658878)	2017
		mCRC, progressed after chemotherapy and with MSI-H and/or dMMR[21]	2017
		Locally advanced or metastatic bladder cancer progressed during or after first-line, adjuvant or neoadjuvant therapy with platinum-based chemotherapy[22]	2017
		SCCHN progressed during or after chemotherapy with a platinum-based drug[23]	2016
		Classical HL relapsed or progressed after autologous HSCT and brentuximab vedotin (Adcetris)[24,25]	2016
		Advanced renal cell carcinoma treated with prior antiangiogenic agents[26]	2015
		Combined with ipilimumab (Yervoy) for unresectable or metastatic melanoma without BRAF V600 mutation[27]	2015
		Advanced nonsquamous NSCLC progressed despite platinum-based chemotherapy[28]	2015
		Advanced squamous NSCLC progressed during or after therapy with platinum-based chemotherapy[29,30]	2015
		Unresectable or metastatic melanoma following ipilimumab and a BRAF inhibitor (in case of BRAF V600 mutation)[31]	2014
Blinatumomab (Blincyto)	Bispecific mAb Targeting CD19 on cancer cells and CD3 on T cells	Adult and pediatric patients with B-cell precursor ALL in first or second complete remission with minimal residual disease greater than or equal to 0.1% (NCT01207388)	2018
		Relapsed or refractory B-cell ALL[32]	2017
		Relapsed or refractory B-cell ALL with negative Philadelphia chromosome[33]	2014

Table 2.1 Immunotherapies Approved by FDA for Cancer—cont'd

Immunotherapy Drug	Category Mechanism of Action	Indication(s)	Year of Approval
Pembrolizumab (Keytruda)	Immune checkpoint inhibitor Blockade of PD-1	Adult and pediatric patients with refractory PMBCL, or who have relapsed after two or more prior lines of therapy (NCT02576990)	2018
		Recurrent or metastatic PD-L1 positive (CPS $\geq$1) cervical cancer with disease progression on or after chemotherapy (NCT02628067)	2018
		Advanced gastric or gastroesophageal junction cancers progressed after $\geq$2 prior lines of standard treatment (NCT02335411)	2017
		Solid tumors in adult and children with MSI-H and/or dMMR, which progressed despite prior treatment and without alternative treatment options[34,35]	2017
		Combined with pemetrexed (Alimta) and carboplatin as first-line treatment for untreated advanced nonsquamous NSCLC, regardless of PD-L1 expression[36]	2017
		Locally advanced or metastatic bladder cancer progressed during or after platinum-based chemotherapy or within 12 months of neoadjuvant or adjuvant chemotherapy, and non eligible for cisplatin-based chemotherapy[37] (NCT02335424)	2017
		Refractory or relapsed classical HL despite $\geq$3 prior treatment[38]	2017
		First-line treatment for metastatic NSCLC with overexpression of PD-L1 and without EGFR or ALK genomic tumor aberrations[39]	2016
		Recurrent or metastatic HNSCC progressed despite chemotherapy (NCT01848834)	2016
		Unresectable or metastatic melanoma[40,41]	2015
		Metastatic NSCLC, expressing PD-L1, and progressed after platinum-based chemotherapy[42]	2015
		Unresectable or metastatic melanoma following ipilimumab and a BRAF inhibitor (in case of BRAF V600 mutation)[43]	2014

Continued

Table 2.1 Immunotherapies Approved by FDA for Cancer—cont'd

Immunotherapy Drug	Category Mechanism of Action	Indication(s)	Year of Approval
Ramucirumab (Cyramza)	mAb Targeting VEGFR2 (KDR)	mCRC progressed after first-line bevacizumab-, oxaliplatin-, and fluoropyrimidine-containing regimen[44]	2015
		Combined with docetaxel for metastatic NSCLC progressed during or after platinum-based chemotherapy and NSCLC with EGFR or ALK alterations progressed during FDA-approved treatment for these aberrations[45]	2014
		Combined with paclitaxel for advanced gastric or gastroesophageal junction adenocarcinoma[46]	2014
		As a single agent for advanced or metastatic, gastric or gastroesophageal junction (GEJ) adenocarcinoma progressed during or after fluoropyrimidine- or platinum-based chemotherapy[47]	2014
Obinutuzumab (Gazyva)	mAb Targeting CD20	Combined with bendamustine followed by obinutuzumab monotherapy in relapsed or refractory follicular lymphoma despite rituximab-containing regimen[48]	2016
		Combined with chlorambucil for untreated CLL[49]	2013
Pertuzumab injection (Perjeta)	mAb Targeting Her2	Combined with trastuzumab and docetaxel as a neoadjuvant for HER2-positive, locally advanced, inflammatory, or early stage breast cancer[50]	2013
		Combined with trastuzumab and docetaxel for HER2-expressing metastatic breast cancer without prior anti-HER2 therapy or chemotherapy[51]	2012
Brentuximab vedotin (Adcetris)	Antibody–drug conjugate Targeting CD30 to deliver a cytotoxic agent	In combination with chemotherapy for adult patients with previously untreated stage III or IV classical HL[52]	2018
		In HL refractory to ASCT or in non candidates for ASCT who received $\geq$2 prior chemotherapy. Systemic anaplastic large cell lymphoma (ALCL) refractory to $\geq$1 prior chemotherapy[53]	2011

Table 2.1 Immunotherapies Approved by FDA for Cancer—cont'd

Immunotherapy Drug	Category Mechanism of Action	Indication(s)	Year of Approval
Ipilimumab (Yervoy)	Immune checkpoint inhibitor Blockade of CTLA-4	In combination with nivolumab for the treatment of patients ≥12 year-old with MSI-H or dMMR mCRC that has progressed following treatment with a fluoropyrimidine, oxaliplatin, and irinotecan.[54]	2018
		Adjuvant treatment of cutaneous melanoma with positive regional lymph nodes of more than 1 mm after complete resection[55]	2015
		Unresectable or metastatic melanoma[56]	2011
Denosumab (Xgeva)	mAb Inhibitor of RANK ligand	Bone giant cell tumor unresectable or with severe morbidity causing resection, in adults and adolescents matured skeletally[57]	2013
		For prevention of skeletal-related events in solid tumors metastases to bone, except for multiple myeloma[58–60]	2010
Sipuleucel-T (Provenge)	DC vaccine Increasing tumor antigen presentation	Hormone-refractory prostate cancer[61]	2010
Ofatumumab (Arzerra)	mAb Targeting CD20	Recurrent or progressive CLL, responded to therapy after ≥2 lines of treatment[62]	2016
		Combined with chlorambucil, for previously untreated cases with CLL inappropriate for fludarabine-based chemotherapy[63]	2014
		Chronic lymphocytic lymphoma refractory to fludarabine and alemtuzumab[64]	2009
Panitumumab (vectibix)	mAb Targeting EGFR	mCRC[65]	2006
Cetuximab (Erbitux)	mAb Targeting EGFR	Combined with FOLFIRI for first-line treatment of mCRC, which is EGFR positive and without K-ras mutation[66]	2012
		Combined with platinum-based chemotherapy plus 5-florouracil (5-FU) for the first-line therapy of recurrent locoregional disease and/or metastatic squamous cell carcinoma of the head and neck[67]	2011
		Combined with radiation therapy for locally or regionally advanced SCCHN	2006

Continued

Table 2.1 Immunotherapies Approved by FDA for Cancer—cont'd

Immunotherapy Drug	Category Mechanism of Action	Indication(s)	Year of Approval
		As a single agent for recurrent or metastatic SCCHN progressed despite platinum-based chemotherapy[68]	
		Combined with irinotecan, for irinotecan-refractory mCRC expressing EGFR[69]	2004
Imiquimod (Aldara)	TLR7 agonist Boosting antitumor immunity	Basal cell carcinoma, actinic keratosis[70]	2004
Bevacizumab (Avastin)	mAb Inhibitor of VEGF-A	In combination with carboplatin and paclitaxel, followed by single-agent bevacizumab for stage III or IV disease after initial surgical resection epithelial ovarian, fallopian tube, or primary peritoneal cancer (NCT00262847)	2018
		Combined with paclitaxel, pegylated liposomal doxorubicin, or topotecan for recurrent epithelial ovarian, fallopian tube, or primary peritoneal cancer resistant to platinum-based chemotherapy[71]	2014
		Combined with paclitaxel and cisplatin or paclitaxel and topotecan for persistent, recurrent or metastatic cervical cancer[72]	2014
		Combined with fluoropyrimidine-based chemotherapy for mCRC progressed during a first-line bevacizumab-based treatment[73]	2013
		Combined with interferon α for mRCC[74]	2009
		As a single agent for glioblastoma, progressed despite prior treatment[75,76]	2009
		Combined with platinum-based chemotherapy as first-line treatment for locally advanced, metastatic, or recurrent NSCLC[77]	2006
		Second-line treatment of mCRC[78]	2006
		Combined with 5-Fluorouracil-based chemotherapy as first-line treatment for mCRC[79]	2004

Table 2.1 Immunotherapies Approved by FDA for Cancer—cont'd

Immunotherapy Drug	Category Mechanism of Action	Indication(s)	Year of Approval
Tositumomab and iodine I-131 tositumomab (Bexxar) (unavailable since Feb 2014)	Radiolabeled mAb Targeting CD20 to deliver radiotherapeutic agent (I-131)	CD-20 positive, follicular NHL, with or without transformation, progresses despite chemotherapy and rituximab[80]	2003
Ibritumomab tiuxetan (Zevalin), Y-90, In-111 ibritumomab	Radiolabeled mAb Targeting CD20 to deliver radiotherapeutic agents	Relapsed or refractory low grade, follicular, or transformed B-cell NHL[81,82]	2002
Alemtuzumab (Campath)[168]	mAb Targeting CD52	CLL[83]	2001, 2007
Gemtuzumab ozogamicin (Mylotarg)	Antibody–drug conjugate Targeting CD33 to deliver a toxic agent	CD33-positive AML in patients above the age of 2 with relapse or not responding to first therapy, recommended with lower dose compared to the first approval[84–86]	2017
		CD33-positive AML (due to sever adverse effects, it was voluntarily removed from the market in 2010)	2000
Ado-trastuzumab emtansine (Kadcyla)	Antibody–drug conjugate Targeting Her2 to deliver a cytotoxic agent	As a single agent for HER2-positive, metastatic breast cancer, progressed despite treatment with trastuzumab and/or a taxane or during or 6 months after completing adjuvant therapy with these agents[87]	2013
Trastuzumab (Herceptin)	mAb Targeting Her2	Combined with cisplatin and a fluoropyrimidine for metastatic gastric or gastroesophageal junction adenocarcinoma overexpressing HER2, not received prior treatment for metastases[88]	2010
		Early stage breast cancer after primary therapy[89]	2006
		HER2-positive breast cancer[90]	1998
Rituxan Hycela (rituximab combined with hyaluronidase)	mAb Targeting CD20	Follicular lymphoma, DLBCL, and CLL[91,92]	2017
Rituximab (Rituxan)	mAb Targeting CD20	Maintenance therapy for previously untreated follicular or B-cell NHL expressing CD20, and responded to rituximab combined with chemotherapy[93]	2011
		Combined with fludarabine and cyclophosphamide for CLL, previously untreated or treated[94]	2010

Continued

Table 2.1 Immunotherapies Approved by FDA for Cancer—cont'd

Immunotherapy Drug	Category Mechanism of Action	Indication(s)	Year of Approval
		First-line treatment for low-grade or follicular, CD20-positive B-cell NHL[95]	2006
		Combined with CHOP or other anthracycline-based chemotherapy regimens as first-line treatment of diffuse large B-cell, CD20-expressing, NHL[96–98]	2006
Rituximab (Rituxan, Mabthera) or 131I-rituximab	Radiolabeled mAb Targeting CD20, delivering radiotherapeutic agents	Relapsed or refractory, CD20-positive, B-cell, low-grade or follicular NHL[99]	1997
Recombinant IL-2 (aldesleukin)	Cytokine Boosting antitumor immunity, growth factor for T cells	Advanced-stage melanoma[100] mRCC[101]	1998 1992
BCG	TLR2 and TLR4 agonists Boosting antitumor immunity	Superficial bladder cancer[102]	1990
IFN-α2B	Cytokine Boosting antitumor immunity	Advanced melanoma[103] Hairy cell leukemia[104]	1996 1986

ALK, *anaplastic lymphoma kinase;* ALL, *acute lymphoblastic leukemia;* AML, *acute myeloid leukemia;* ASCT, *autologous stem cell transplant;* BCG, *Bacillus Calmette–Guérin;* CAR, *chimeric antigen receptor;* CHOP, *cyclophosphamide, doxorubicin, vincristine, and prednisone;* CLL, *chronic lymphocytic leukemia;* CPS, *combined positive score;* CTLA-4, *cytotoxic T-lymphocyte–associated antigen 4;* DLBCL, *diffuse large B-cell lymphoma;* dMMR, *DNA mismatch repair deficiency;* EGFR, *epidermal growth factor receptor;* FDA, *Food and Drug Administration;* GD2, *disialoganglioside GD2;* GM-CSF, *granulocyte macrophage colony–stimulating factor;* Her2, *human epidermal growth factor receptor 2;* HL, *Hodgkin lymphoma;* HSCT, *hematopoietic stem cell transplantation;* IFN, *interferon;* IL-2, *interleukin-2;* mAb, *monoclonal antibody;* mCRC, *metastatic colorectal cancer;* mRCC, *metastatic renal cell carcinoma;* MSI-H, *high microsatellite instability;* NHL, *non-Hodgkin lymphoma;* NSCLC, *non-small cell lung cancer;* PDGF, *platelet-derived growth factor;* PD-L1, *programmed death-ligand 1;* PMBCL, *primary mediastinal large B-cell lymphoma;* RANK, *receptor activator of nuclear factor kappa-B;* SCCHN, *squamous cell cancer of the head and neck;* SLAMF7, *signaling lymphocytic activation molecule F7;* TLR, *toll-like receptor;* VEGF-A, *vascular endothelial growth factor-A;* VEGFR2, *vascular endothelial growth factor receptor-2.*

PASSIVE VERSUS ACTIVE IMMUNOTHERAPY

Cancer immunotherapy generally has two main arms: active and passive. The mechanism of action of the therapeutic and the immune system condition of the patient provide the basis for this classification. In passive immunotherapy, the cells activated *ex vivo* or the molecules expressed in low levels in the body rectify the defective immune responses. They are usually taken into consideration for those who have impotent immune systems against cancer. The main passive immunotherapeutics to date consist of monoclonal antibodies, adoptive transfer of the immune cells preactivated *ex vivo*, adjuvants, and recombinant cytokines. Passive immunotherapy has shown short-term effects compared to active type, and it might need more repetitions of administration to be efficient.[105,106]

Active immunotherapy is designed to activate effector functions of the immune system *in vivo*. The status of the immune system should be suitable for active immunotherapy. In this case, the immune system is responsive upon challenge, becomes stimulated properly, and executes the effector functions. In the active immunotherapy, activation of an endogenous and long-lasting immune response is the main goal. The most important active approaches include various tumor vaccines (peptide vaccines, whole tumor cells, dendritic cells (DCs) or viruses as vehicles for delivery of the tumor antigen or its genetic material), blockade of checkpoint inhibitors of T cell activation, and oncolytic viruses.[105,106]

PASSIVE IMMUNOTHERAPY FOR CANCER

Cytokines

Cytokines are among the immunotherapeutic modalities used for cancer treatment. Among them IFN-α, IL-2, and IL-12 have been used for a long time. However, compared to the past, their combination therapy excels to monotherapy at this time.

IFN-α has various antitumor activities categorized as tumor-based and immune cell–based effects. In tumor cells, it hampers cell growth, and incudes apoptosis, cell differentiation, and migration and increases cell surface antigen expression. In discussion of the immune system–dependent effects of interferons, increase in activation and effector function of DCs, αβ and γδ T cells, and inhibitory effect on immune-suppressing immune cells such as Tregs and MDSCs are mentioned.[107]

IL-2, firstly exploited as a T cell growth factor for cancer immunotherapy, showed severe toxicity in high doses.[108] It also has dual effects on immune cells, and it is capable of expanding both regulatory T cells (Tregs) and effectors, which make this cytokine a candidate for immunotherapy of both cancer and autoimmune diseases as well.[109] IL-2, which is among the first immunotherapeutics approved as monotherapy, is not standardly administered alone nowadays. Combined with other therapeutics, including peptide vaccines, adoptive T cell therapy (ACT), immune checkpoint inhibitors, other cytokines, mAbs chemotherapy, and targeted therapy, it showed superior effectiveness.[110]

IL-12 is the other cytokine potential for cancer immunotherapy. This cytokine, which is produced by antigen presenting cells (APCs) facing antigen stimulation in normal condition, polarizes CD4+ T cell to Th1 cells, and increases CD8+ T cells, natural killer (NK) and natural killer T cell (NKT cells) and induces Treg cells apoptosis.[111]

Monoclonal Antibodies (mAbs)

Monoclonal antibodies (mAbs) are the very historical "magical bullets." At the beginning of their development, murine mAbs were used, which caused a big problem as they were recognized as non-self-proteins and were attacked by secretion of human antibodies against the mAbs, or the human anti-mouse antibody (HAMA), which counteracted their efficacy. Antibody engineering diminished the problem with generation of chimeric, humanized, or fully human mAbs.[112]

MAbs target a variety of antigens including tumor antigens and cell surface molecules initiating signaling pathways. In case of blocking tumor-specific antigens, one or more of the following mechanisms might be used for destroying the tumor cells: antibody-dependent cellular cytotoxicity (ADCC), complement-dependent cytotoxicity (CDC), and antibody-dependent cellular phagocytosis (ADCP). MAbs having constant fragment (FC) induce cytotoxic effects through ligation with their Fcγ receptors on the surface of immune cells. FcγRIIIa on the surface of NK cells mediates ADCC, and FcγRIIa on the surface of macrophages operates ADCP. MAbs used to block cell surface receptors routinely block a signaling pathway and its downstream necessary for tumor progress or invasiveness.[113,114] MAbs are capable of targeting various cell markers and hamper their subsequent events. Inhibiting targets that play a role in angiogenesis or immunosuppression has been achieved by mAbs as well. Targeting vascular endothelial growth factor (VEGF) and CD25, a cell marker of Tregs, are the examples, respectively.[115,116]

Antibodies are able to carry cytotoxic agents linked to their Fc and deliver them to the target site. Toxins, chemotherapeutics, radioisotopes, cytokines, and enzymes convert prodrugs to cytotoxic drugs.[117–119]

The new version of antibodies, named as bispecific mAbs, target two antigen/receptors on the surface of one or two kinds of cells. For example, they can involve both tumor cells and T cells without link between T cell receptor (TCR) and peptide/MHC complex, because they contain two single-chain variable fragments (scFvs) specific for tumor antigen and the T cells marker, CD3.[120]

Adoptive Cell Therapy

In adoptive cell therapy (ACT), *ex vivo* expanded and activated autologous or allogeneic immune cells, such as NK cells and T cells, are transferred to the patient.[121–123] Tumor infiltrating lymphocytes (TILs) are isolated from the patients' own resected melanoma lesions or the donor, cocultured with high-dose IL-2 *in vitro* and then are examined for specificity to the tumor antigen and transferred to the patients according to different schedules.[124]

Cytokine-induced killer (CIK) cells are used for ACT as well. They are an advanced type of lymphokine activated killer (LAK) cell. Both are heterogeneous immune cells including CD3+CD56+, CD3+CD56−, and CD3−CD56+. The majority of LAK cells are NK cells whereas CIK cells mostly consist of NKT cells, CD3+CD56+ cells, extracted from peripheral blood mononuclear cells. LAK cells are cocultured with IL-2, whereas CIK cells are cocultured with anti-CD3, IL-2, IL-1, IFN-γ, and other cytokines such as combination of IL-2 and IL-15, which can enhance their cytotoxic activity.[125–127] They account for non-major histocompatibility (MHC)-restricted immunity against tumor, which makes this approach appropriate for cases with reduction in antigen or MHC expression or allogenic tumors. Their cytotoxic activity depends on perforin and Fas/FasL.[128]

Cascade-primed (CAPRI) cell therapy is another type of ACT, in which peripheral blood monocytes from the patient are extracted and applied for activating naïve T cells into cytotoxic phenotypes because they are a good source of tumor antigens of the patient. OKT3, anti-CD3 antibody, and IL-2 are used for culturing T cells, which then activate monocytes to present more tumor antigens. These monocytes are responsible for priming the other naïve T cells and, as a result, making the CAPRI cells.[129]

Macrophage-activated killer (MAK) and MAK-dendritic cell (MAK-DC) are two other kinds of ACT using macrophages and DCs for adoptive immunotherapy.[127]

Among the adoptive immune cells transfer approaches, T cell receptors can show more avidity for tumor antigens through engineering. Chimeric antigen receptors (CARs) are engineered TCRs, whose ectodomain contains the scFv from an antibody matching a tumor antigen and the endodomain has CD3ζ chain, similar to the TCR, and different costimulatory molecules have been added in each new generation.[130]

The fourth generation of CARs are T cells redirected for universal cytokine-mediated killing (TRUCKs). They carry IL-12 vector, which unleashes IL-12 by connection of the tumor antigen to the CAR.[131] It is a novel approach to avoid systemic effects of cytokine administration. In this case, IL-12 makes a local inflammation and recruits innate immunity for tumor elimination.[132]

ACTIVE IMMUNOTHERAPY FOR CANCER

Vaccines

Vaccines for cancer treatment are focused on activate immunity against tumor-specific or tumor-associated antigens *in vivo*. The variety of tumor antigens explained previously are targets of vaccination. There are different approaches and vectors to deliver tumor antigens to the patient and augment antitumor immunity. These approaches shape the basis for cancer vaccine classification.[130] Here we briefly review four main classes of therapeutic cancer vaccines.

Peptide Vaccine

In this type, immunogenic epitopes of tumor antigens are generated and delivered into the patient. Therefore, identification of specific and immunogenic tumor antigens is crucial to induce a robust antitumor response. Early on, peptide vaccines were being applied as monotherapy using limited epitopes, which resulted in low efficacy because of the confined *in vivo* immune responses and MHC restriction. Since the epitopes are presented with special MHCs, knowledge of the patient's human leukocyte antigen (HLA) is required for peptide vaccination.[133] To overcome MHC restriction, both cytotoxic T lymphocyte (CTL) and helper T (Th) cell epitopes are used in combination as peptide cocktail or to produce synthetic peptides, which are longer and more immunogenic as well.[134]

Tumor Cell Vaccine

Vaccination by autologous tumor cells has several benefits as they are a great source of tumor antigens. This type of vaccine contains the antigens that might activate both Th cells and CTLs and might be individualized for the patient. Therefore, they are considered as personalized vaccines. However, if the tumor does not present enough antigens, the vaccine efficacy will be limited.[135]

Using allogeneic whole tumor cells for vaccination has also several advantages, as they are immortal cell lines, with simple and standard production protocols, which make them cost-effective, good sources of tumor antigens and allow comparing the clinical responses between patients.[136]

To improve immunogenicity of the allogeneic whole tumor cells, they are transfected by the genes of cytokines or costimulatory molecules. Then, they are inactivated by irradiation and transferred to the body as the tumor antigen source. One of the common whole cell vaccines, applicable for a variety of tumors, is designed to secrete granulocyte-macrophage colony-stimulating factor (GM-CSF). The cytokine is capable of activating APCs and enhancing cross-presentation of the antigens carried by the vaccine. This vaccine is generally known as GVAX.[137]

Dendritic Cell Vaccine

DCs are professional APCs, which can be loaded with candidate tumor antigens *ex vivo* or *in vivo*. In the first method, DCs are derived from peripheral blood monocytes and are cocultured with the tumor antigen *ex vivo* plus cytokines such as IL-4 and GM-SCF necessary for DC maturation, which are infused to the patient and result in T cell activation.[138]

The tumor DNA, mRNA, and viruses, as vectors, are the other options for generating DC vaccine *ex vivo*. They usually encode a specific tumor antigen.[139–141] Autologous whole tumor cells can be also used instead of peptides to prepare DC vaccine since they are a source of all identified and unidentified tumor antigens.[142]

Providing the DCs with tumor antigens *in vivo* is carried out through using antibodies, which target a receptor on the surface of DCs and, in addition, are linked to a specific tumor antigen. There are some points for both methods that should be taken into consideration: selection of DC subsets, and providing maturation signals and costimulatory molecules.[143]

Genetic Vaccine

The DNA encoding a specific tumor antigen or the specified mRNA is infused to patients as genetic vaccine. The genetic material must be obtained and used for tumor antigen production by host cells, and then the antigen must be processed and cross-presented by APCs, which activate T cells. Since there are some practical limitations in the mentioned events, optimization is required to elicit strong immune responses against tumor.[144] Delivery systems and vehicles, adjuvants, costimulatory molecules, and route of administration can be modified to improve the vaccine efficacy.[145]

One of vectors used for their delivery are viruses such as adenovirus and vaccinia virus. They are easily entered to the host cells due to the infectivity potential, although development of antibodies to neutralize these viruses decreases their efficacy in the next injections. An approach has been developed to overcome this obstacle. Using heterologous vectors, which share the same antigen, for priming (first dose) and boosting (next doses), prevents production of antibodies that reduce the efficacy of vaccine.[146]

Checkpoint Inhibitors

As explained previously, two signals are required for T cell activation: the first is done by connection of peptide/MHC complex and TCR, and the second must be provided through costimulatory molecules. The CD28 molecule on the surface of T cells binds to CD80 (B7-1) and CD86 (B7-2) on the surface of APCs to create the costimulatory signal. CTLA-4, a molecule on the surface of T cells, competes with CD28 to engage B7 molecules. CTLA-4 is a rival for CD28 and inhibits T cell activation. Using antibodies that block CTLA-4 leads to hampering its inhibitory effects and increases antitumor and clinical responses.[147]

Programmed cell death protein 1 (PD-1) is another coinhibitory molecule, which is expressed on T cells in the late phase of their effector function, contrary to CTLA-4. It has two ligands, PD-L1 and PD-L2, to which binding results in inhibiting T cell proliferation and activity. This is also responsible for inducing T cell apoptosis. Increased expression of the ligands has been reported in numerous malignancies and mAbs blocking PD-1 or its ligands have been applied as immunotherapy for cancer and has shown promising results.[148]

Oncolytic Viruses

There are viruses that preferentially replicate in tumor cells and destroy them. Tumor antigens are disseminated following the tumor cell lysis and cell-mediated immunity is activated. The replicated oncolytic viruses are subsequently released to infect and kill more tumor cells. The immune system might also develop antibodies to fight the viral infection. Normal cells use mechanisms to omit viruses. However, these defective mechanisms in tumor cells cease the virus clearance.[149] DNA oncolytic viruses can be engineered to increase their lytic properties and immune activating functions.[150] Herpes simplex virus type 1 for melanoma, vaccinia virus JX-594 for hepatocellular carcinoma, adenovirus for bladder cancer, and reovirus for head and neck cancer have been applied as oncolytic viruses.[151]

MECHANISM-BASED IMMUNOTHERAPIES: INTERACTIONS BETWEEN TUMOR CELLS AND THE IMMUNE SYSTEM

The dynamic interaction between malignant cells and T cell-mediated immunity have been reviewed, and various stages appeared to have potential for interference and breaking the pathological cycle. From release of tumor antigens to cytotoxic effects of activated T cells, there are four main stages that can be considered for intervention. In the following section, we will discuss approaches to empower these stages and evoke the physiological function of T cells.

TUMOR ANTIGEN EXPRESSION, RELEASE AND PRESENTATION

To recognize and destroy tumor cells, the first step is release of tumor antigens and their presentation by APCs. Therapies that destroy tumor cells can accelerate antigen release. Oncolytic virus described recently is the immunotherapeutic approach with this aim. Other conventional therapies perform this action as well. Chemotherapy, which has immunomodulatory effects, acts by two main ways: (1) induction of immunogenic cell death, and (2) decreasing tumor bulk and inhibiting tumor immune escape.[152]

Radiotherapy, in addition to killing tumor cells, results in tumor antigen distribution, enhancing the expression of MHC class I on APCs via three paths: (1) and (2) augmentation of intracellular peptide pool by unfolding and degradation of proteins and by increase in protein synthesis, and (3) production of radiation-specific antigens leading to expansion of peptide pool diversity.[153]

There are also oncogenic signaling pathways that can be inhibited by targeted therapies. The treatment shows some immunologic alterations. For example, inhibition of MEK/BRAF in patients with melanoma led to upregulation of tumor differentiation antigens and MHC molecules and some other changes in cytokine secretion and T cell activation favoring antitumor immunity.[154]

Vaccination is the other modality to increase tumor antigens and their presentation via APCs. As explained, various vaccines are designed to deliver tumor antigens to APCs through different ways. Tumor antigen peptides can be directly delivered through various types of vaccines. The DNA or RNA encoding the tumor antigen can also be delivered, with or without viral vehicles, as genetic vaccines. In the more complex type of vaccine, APCs are extracted from peripheral blood monocytes and loaded with the tumor antigen or the related DNA or RNA and afterward injected to the patient.[130]

Cytokines, administered alone or as adjuvants for vaccines, also increase tumor antigen presentation. For example, GM-CSF plays a role in expansion, activation, maturation, and recruitment of DCs as well as upregulation of CD80 and CD1d on DCs to induce robust immune responses.[155] Type I IFNs show effects on regulation of the functions of almost all immune cells and coordinate antitumor immune responses. To release and present tumor antigens efficiently, it induces activation of DCs for cross-presentation of tumor antigen to T cells and high expression of tumor antigens.[107]

Toll-like receptor (TLR) agonists are the other components inducing "danger signal" to awake immune responses and cease tolerance against tumor cells. They are used alone or in combination with other therapies such using them as adjuvants for vaccines. Coley toxin and Bacillus Calmette–Guerin (BCG) are TLR2 and TLR4 agonists, which have been used for long time as cancer treatment. Overall, they are microbial components that activate innate immune system through TLRs, which subsequently causes an increase in expression of costimulatory molecules (CD80, CD86, and CD40), DC maturation, and antigen presentation. TLR agonists are also capable of increasing inflammatory cytokines such as IL-12 and tumor necrosis factor (TNF)-a, which induce adaptive immunity against tumor. There are also immune inhibitory effects for TLR agonists, which are not relevant to the topic at hand.[156,157]

CD40 molecule is a member of the tumor necrosis factor (TNF) receptor superfamily, expressed on the APCs including DCs, monocytes, B cells, and some tumor cells. Its natural ligands on T helper cells and agonistic mAbs targeting CD40 give rise to the APC activation and antitumor immunity. Increasing myeloid cells to fight cancer has been observed in combination of chemotherapy and agonistic CD40 mAb.[158,159]

T CELL PRIMING AND ACTIVATION

After cross-presentation of the tumor antigen, CD8+ T cell activation, or cross-priming, occurs. There must be adequate quantity of T cells and sufficient costimulatory signals besides activated APCs and peptide/MHC complexes engaging TCR to prime and activate T cells. Therefore, cytokines required for T cell proliferation and activation and costimulatory molecules propelling T cell activation can be manipulated to result in a direction toward antitumor immunity.[160,161]

In addition to immune checkpoints inhibiting T cell activation such as CTLA-4, there are molecules on T cells that are immune checkpoints stimulating T cell activation like CD137, OX-40, and CD27. Monoclonal antibodies have been designed to target these immune checkpoints with blocking effects on inhibitory immune checkpoints and agonistic actions targeting costimulatory molecules.[162–165]

CD137 or 4-1BB, OX-40, and CD27 are members of TNF receptor super family expressed on T cells and by binding to their ligand on APCs this results in T cell activation. Various immunotherapeutic modalities target costimulatory molecules to exploit this property. Agonistic mAbs, bispecific Abs, and CAR T cells are the examples aimed to use these molecules to increase expansion, activation, and survival of T cells.[163–166]

Cytokines have an influence on the expansion and activation of T cells as well. IL-2 is among the best-known cytokines, which has been used as immunotherapy and has shown clinical responses and tumor regression. CD4+ T cells are the cellular source of this cytokine. It is also applied for *in vitro* expansion of T cells in ACT method.[167,168] IL-12 is the next important cytokine in T cell proliferation and activation. It is secreted from APCs and connects the innate and adaptive immunity in inflammatory conditions. IL-12 induces Th1 polarization and gives rise to the level of other cytokines including IFN-γ and IL-2. Taken together, this cytokine plays several roles in various stages and affects immune cells; however, orchestration of Th1 polarization to attack pathogens and tumor cells is its most prominent role.[169,170]

TRAFFICKING AND INFILTRATION OF T CELLS TO TUMOR

T cells must reach the tumor microenvironment to exert their antitumor functions and advance elimination of tumor cells. To provide this access, chemokine receptors are set in the tumor microenvironment and act by chemokine engagement. In fact, in the tumor microenvironment chemokines secreted by immune and nonimmune (tumor and nontumor) cells attract immune cells for attacking the tumor cells.[161,171]

TILs express a number of chemokine receptors playing roles in chemotaxis, for instance, CXCR3, CCR5, and CX3CR1. Increased expression of their related ligands brings about infiltration of T cells to the tumor microenvironment. These ligands are CXCL9, and CXCL10 binding to CXCR3, CCL5 binding to CCR5, and CX3CL1 binding to CX3CR1. CXCL16 is another ligand, which binds to CXCR6 and exhibits chemoattractant feature. Taken together, elevated production of chemokines such as CXCL9, CXCL10, CCL5, CX3CL1, and CXCL16 promote infiltration of T cells from lymph nodes to the tumor site.[171,172]

VEGF, which is a known angiogenic factor, has displayed various negative effects on immune cells during antitumor activity as well. Its important inhibitory functions are performed to hamper maturation of DCs and trafficking of T cells to the tumor site. Applying modalities, e.g., mAbs, to block its activity showed increased tumor infiltration of T cells and led to improved clinical outcomes. Other angiogenic factors such as angiopoietins hinder immune cell trafficking by changing the expression level of adhesion molecules.[173,174]

TUMOR CELL RECOGNITION AND EFFECTOR FUNCTION OF T CELLS

To enhance tumor antigen recognition, CARs, the antigen receptors containing both scFv from an antibody targeting tumor antigen and signaling domains from T cell, have been designed. Costimulatory molecules have additively been installed in each generation of CARs to improve their function. To express CARs, T cells are genetically engineered via transfection by viral vehicles, frequently a retrovirus or a lentivirus delivering the encoding material, and T cells are proliferated and transferred to the patient adoptively.[175,176]

As reviewed in the elimination phase of tumor immunoediting, IFN-γ exhibits some antitumor functions including cytotoxic effects and expression of MHC-I. However, application of cytokines alone is not of interest these days. Recently, a new approach has been investigated to use genetically modified bacteria encoding IFN-γ, which preferentially grows in tumor cells, for cancer treatment.[177]

In contrast to CTLA-4, which plays a role in the activation of T cells in early stages and in the secondary lymphoid tissues, PD-1 and its ligands, PD-L1 (B7H1) and PD-L2 (B7-DC), act during the T cell effector functions. They usually play a role in inhibition of T cell effector functions in peripheral tissues such as tumor microenvironment and create an immunosuppressive state there.[178] There are also other mechanisms for their inhibitory action such as inducing apoptosis of activated T cells, T cell anergy and exhaustion, proliferation of Treg cells, and blocking T cell proliferation and activation.[179] Antibodies blocking PD-1 or PD-L1 have been tested to a great extent for various advanced cancer types and displayed effectiveness.[180]

There are other immune inhibitory checkpoints, whose blockade counteracts their negative effects. T cell immunoglobulin and mucin domain 3 (TIM-3) is one example, which leads to exhaustion of tumor antigen-specific CD8+ T cells, and its administration in combination with PD-1/PD-L1 blocking antibodies is being investigated in clinical trials.[181,182] Lymphocyte-activation gene 3 is also the next immune checkpoint inhibitor that is being evaluated in clinical trials.[130]

Indoleamine 2,3-dioxygenase (IDO) is an enzyme that acts as an endogenous pathway to induce immunosuppressive tumor microenvironment. IDO is mostly secreted by tumor cells and some immune cells, e.g., macrophages and DCs.[183,184] IDO is the catalyzer in tryptophan degradation process. Reduced levels of tryptophan inhibits T cell proliferation and production of kynurenine, and the bioactive tryptophan metabolite induces T cell apoptosis as well as inhibiting T cell proliferation.[185–187] IDO showed other suppressive effects on immunity such as inhibiting NK cell proliferation, increasing immune suppressive cells, e.g., Tregs and MDSC, and induction of tumor angiogenesis.[188] It has also been found as one of the main performers in resistance to the anti-CTLA-4 immune checkpoint inhibitor, ipilimumab. In order to counteract IDO's activity, small-molecule inhibitors are being investigated for cancer treatment in clinical trials. It has also been targeted by a peptide vaccine in a recent trial.[189,190]

REFERENCES

1. Coley II WB. Contribution to the knowledge of sarcoma. *Ann Surg* 1891;**14**(3):199–220.
2. McCarthy EF. The toxins of William B. Coley and the treatment of Bone and soft-tissue sarcomas. *Iowa Orthop J* 2006;**26**:154–8.
3. Kasten FH. Paul Ehrlich: pathfinder in cell biology. 1. Chronicle of his life and accomplishments in immunology, cancer research, and chemotherapy. *Biotech Histochem* 1996;**71**(1):2–37.
4. Dunn GP, Old LJ, Schreiber RD. The immunobiology of cancer immunosurveillance and immunoediting. *Immunity* 2004;**21**(2):137–48.
5. Kochenderfer JN, et al. Lymphoma remissions caused by anti-CD19 chimeric antigen receptor T cells are associated with high serum Interleukin-15 levels. *J Clin Oncol* 2017;**35**(16):1803–13.
6. Kantarjian HM, et al. Inotuzumab ozogamicin versus standard therapy for acute lymphoblastic leukemia. *N Engl J Med* 2016;**375**(8):740–53.
7. Antonia SJ, et al. Durvalumab after chemoradiotherapy in stage III non-small-cell lung cancer. *N Engl J Med* 2017;**377**(20):1919–29.
8. Powles T, et al. Efficacy and safety of durvalumab in locally advanced or metastatic urothelial carcinoma: updated results from a phase 1/2 open-label study. *JAMA Oncol* 2017;**3**(9):e172411.
9. Apolo AB, et al. Updated efficacy and safety of avelumab in metastatic urothelial carcinoma (mUC): pooled analysis from 2 cohorts of the phase 1b Javelin solid tumor study. *J Clin Oncol* 2017;**35**(15 Suppl.):4528.
10. Kaufman HL, et al. Avelumab in patients with chemotherapy-refractory metastatic Merkel cell carcinoma: a multicentre, single-group, open-label, phase 2 trial. *Lancet Oncol* 2016;**17**(10):1374–85.
11. Tap WD, et al. Olaratumab and doxorubicin versus doxorubicin alone for treatment of soft-tissue sarcoma: an open-label phase 1b and randomised phase 2 trial. *Lancet* 2016;**388**(10043):488–97.
12. Fehrenbacher L, et al. Atezolizumab versus docetaxel for patients with previously treated non-small-cell lung cancer (POPLAR): a multicentre, open-label, phase 2 randomised controlled trial. *Lancet* 2016; **387**(10030):1837–46.
13. Rosenberg JE. Atezolizumab in patients with locally advanced and metastatic urothelial carcinoma who have progressed following treatment with platinum-based chemotherapy: a single-arm, multicentre, phase 2 trial. *Lancet* 2016:387.
14. Lonial S, et al. Elotuzumab therapy for relapsed or refractory multiple myeloma. *N Engl J Med* 2015;**373**(7): 621–31.
15. Thatcher N, et al. Necitumumab plus gemcitabine and cisplatin versus gemcitabine and cisplatin alone as first-line therapy in patients with stage IV squamous non-small-cell lung cancer (SQUIRE): an open-label, randomised, controlled phase 3 trial. *Lancet Oncol* 2015;**16**(7):763–74.
16. Dimopoulos MA, et al. Daratumumab, lenalidomide, and dexamethasone for multiple myeloma. *N Engl J Med* 2016;**375**(14):1319–31.
17. Palumbo A, et al. Daratumumab, bortezomib, and dexamethasone for multiple myeloma. *N Engl J Med* 2016;**375**(8):754–66.

18. Usmani S, et al. Clinical efficacy of daratumumab monotherapy in patients with heavily pretreated relapsed or refractory multiple myeloma. *Blood* 2015;**126**(23):29.
19. Andtbacka RH, et al. Talimogene laherparepvec improves durable response rate in patients with advanced melanoma. *J Clin Oncol* 2015;**33**(25):2780–8.
20. Yu AL, et al. Anti-GD2 antibody with GM-CSF, Interleukin-2, and isotretinoin for neuroblastoma. *N Engl J Med* 2010;**363**(14):1324–34.
21. Overman MJ, et al. Nivolumab in patients with metastatic DNA mismatch repair-deficient or microsatellite instability-high colorectal cancer (CheckMate 142): an open-label, multicentre, phase 2 study. *Lancet Oncol* 2017;**18**(9):1182–91.
22. Sharma P, et al. Nivolumab in metastatic urothelial carcinoma after platinum therapy (CheckMate 275): a multicentre, single-arm, phase 2 trial. *Lancet Oncol* 2017;**18**(3):312–22.
23. Ferris RL, et al. Nivolumab for recurrent squamous-cell carcinoma of the head and neck. *N Engl J Med* 2016;**375**(19):1856–67.
24. Ansell SM, et al. PD-1 Blockade with nivolumab in relapsed or refractory Hodgkin's lymphoma. *N Engl J Med* 2015;**372**(4):311–9.
25. Younes A, et al. Nivolumab for classical Hodgkin's lymphoma after failure of both autologous stem-cell transplantation and brentuximab vedotin: a multicentre, multicohort, single-arm phase 2 trial. *Lancet Oncol* 2016;**17**(9):1283–94.
26. Motzer RJ. Nivolumab versus everolimus in advanced renal-cell carcinoma. *N Engl J Med* 2015:373.
27. Postow MA, et al. Nivolumab and ipilimumab versus ipilimumab in untreated melanoma. *N Engl J Med* 2015;**372**(21):2006–17.
28. Borghaei H. Nivolumab versus docetaxel in advanced nonsquamous non-small-cell lung cancer. *N Engl J Med* 2015;**373**.
29. Brahmer J. Nivolumab versus docetaxel in advanced squamous-cell non-small-cell lung cancer. *N Engl J Med* 2015;**373**.
30. Rizvi NA, et al. Activity and safety of nivolumab, an anti-PD-1 immune checkpoint inhibitor, for patients with advanced, refractory squamous non-small-cell lung cancer (CheckMate 063): a phase 2, single-arm trial. *Lancet Oncol* 2015;**16**(3):257–65.
31. Weber JS. Nivolumab versus chemotherapy in patients with advanced melanoma who progressed after anti-CTLA-4 treatment (CheckMate 037): a randomised, controlled, open-label, phase 3 trial. *Lancet Oncol* 2015:16.
32. Kantarjian H, et al. Blinatumomab versus chemotherapy for advanced acute lymphoblastic leukemia. *N Engl J Med* 2017;**376**(9):836–47.
33. Topp MS, et al. Safety and activity of blinatumomab for adult patients with relapsed or refractory B-precursor acute lymphoblastic leukaemia: a multicentre, single-arm, phase 2 study. *Lancet Oncol* 2015; **16**(1):57–66.
34. Le DT. PD-1 Blockade in tumors with mismatch-repair deficiency. *N Engl J Med* 2015:372.
35. Le DT, Durham JN. Mismatch repair deficiency predicts response of solid tumors to PD-1 blockade. *Science* 2017;**357**(6349):409–13.
36. Langer CJ, et al. Carboplatin and pemetrexed with or without pembrolizumab for advanced, non-squamous non-small-cell lung cancer: a randomised, phase 2 cohort of the open-label KEYNOTE-021 study. *Lancet Oncol* 2016;**17**(11):1497–508.
37. Bellmunt J, et al. Pembrolizumab as second-line therapy for advanced urothelial carcinoma. *N Engl J Med* 2017;**376**(11):1015–26.
38. Chen R, et al. Phase II study of the efficacy and safety of pembrolizumab for relapsed/refractory classic hodgkin lymphoma. *J Clin Oncol* 2017;**35**(19):2125–32.

39. Reck M, et al. Pembrolizumab versus chemotherapy for PD-L1−positive non−small-cell lung cancer. *N Engl J Med* 2016;**375**(19):1823−33.
40. Robert C, et al. Pembrolizumab versus ipilimumab in advanced melanoma. *N Engl J Med* 2015;**372**(26): 2521−32.
41. Ribas A. Pembrolizumab versus investigator-choice chemotherapy for ipilimumab-refractory melanoma (KEYNOTE-002): a randomised, controlled, phase 2 trial. *Lancet Oncol* 2015:16.
42. Garon EB. Pembrolizumab for the treatment of non-small-cell lung cancer. *N Engl J Med* 2015:372.
43. Robert C, et al. Anti-programmed-death-receptor-1 treatment with pembrolizumab in ipilimumab-refractory advanced melanoma: a randomised dose-comparison cohort of a phase 1 trial. *Lancet* 2014; **384**(9948):1109−17.
44. Tabernero J, et al. Ramucirumab versus placebo in combination with second-line FOLFIRI in patients with metastatic colorectal carcinoma that progressed during or after first-line therapy with bevacizumab, oxaliplatin, and a fluoropyrimidine (RAISE): a randomised, double-blind, multicentre, phase 3 study. *Lancet Oncol* 2015;**16**(5):499−508.
45. Garon EB, et al. Ramucirumab plus docetaxel versus placebo plus docetaxel for second-line treatment of stage IV non-small-cell lung cancer after disease progression on platinum-based therapy (REVEL): a multicentre, double-blind, randomised phase 3 trial. *Lancet* 2014;**384**(9944):665−73.
46. Wilke H, et al. Ramucirumab plus paclitaxel versus placebo plus paclitaxel in patients with previously treated advanced gastric or gastro-oesophageal junction adenocarcinoma (RAINBOW): a double-blind, randomised phase 3 trial. *Lancet Oncol* 2014;**15**(11):1224−35.
47. Fuchs CS, et al. Ramucirumab monotherapy for previously treated advanced gastric or gastro-oesophageal junction adenocarcinoma (REGARD): an international, randomised, multicentre, placebo-controlled, phase 3 trial. *Lancet* 2014;**383**(9911):31−9.
48. Sehn LH, et al. Obinutuzumab plus bendamustine versus bendamustine monotherapy in patients with rituximab-refractory indolent non-Hodgkin lymphoma (GADOLIN): a randomised, controlled, open-label, multicentre, phase 3 trial. *Lancet Oncol* 2016;**17**(8):1081−93.
49. Goede V, et al. Obinutuzumab (GA101) plus chlorambucil (Clb) or rituximab (R) plus Clb versus Clb alone in patients with chronic lymphocytic leukemia (CLL) and preexisting medical conditions (comorbidities): final stage 1 results of the CLL11 (BO21004) phase III trial. *J Clin Oncol* 2013;**31**(15 Suppl.):7004.
50. Gianni L, et al. Efficacy and safety of neoadjuvant pertuzumab and trastuzumab in women with locally advanced, inflammatory, or early HER2-positive breast cancer (NeoSphere): a randomised multicentre, open-label, phase 2 trial. *Lancet Oncol* 2012;**13**(1):25−32.
51. Baselga J, et al. Pertuzumab plus trastuzumab plus docetaxel for metastatic Breast cancer. *N Engl J Med* 2012;**366**(2):109−19.
52. Connors JM, et al. Brentuximab vedotin with chemotherapy for stage III or IV Hodgkin's lymphoma. *N Engl J Med* 2018;**378**(4):331−44.
53. Younes A, et al. Results of a pivotal phase II study of Brentuximab vedotin for patients with relapsed or refractory Hodgkin's lymphoma. *J Clin Oncol* 2012;**30**(18):2183−9.
54. Overman MJ, et al. Durable Clinical Benefit With Nivolumab Plus Ipilimumab in DNA Mismatch Repair-Deficient/Microsatellite Instability-High Metastatic Colorectal Cancer. *J Clin Oncol* 2018;**36**(8):773−9.
55. Eggermont AMM, et al. Adjuvant ipilimumab versus placebo after complete resection of high-risk stage III melanoma (EORTC 18071): a randomised, double-blind, phase 3 trial. *Lancet Oncol* 2015;**16**(5):522−30.
56. Hodi FS, et al. Improved survival with ipilimumab in patients with metastatic melanoma. *N Engl J Med* 2010;**363**(8):711−23.
57. Chawla S, et al. Safety and efficacy of denosumab for adults and skeletally mature adolescents with giant cell tumour of bone: interim analysis of an open-label, parallel-group, phase 2 study. *Lancet Oncol* 2013; **14**(9):901−8.

58. Fizazi K, et al. A randomized phase III trial of denosumab versus zoledronic acid in patients with bone metastases from castration-resistant prostate cancer. *J Clin Oncol* 2010;**28**(18 Suppl.):LBA4507.
59. Stopeck A, et al. Abstract P6-14-01: effect of denosumab versus zoledronic acid treatment in patients with Breast cancer and Bone metastases: results from the extended blinded treatment phase. *Cancer Res* 2010; **70**(24 Suppl.):P6-14-01.
60. Henry DH, et al. Randomized, double-Blind study of denosumab versus zoledronic acid in the treatment of Bone metastases in patients with advanced cancer (excluding Breast and prostate cancer) or multiple myeloma. *J Clin Oncol* 2011;**29**(9):1125–32.
61. Kantoff PW, et al. Sipuleucel-t immunotherapy for castration-resistant prostate cancer. *N Engl J Med* 2010; **363**(5):411–22.
62. van Oers MHJ, et al. Ofatumumab maintenance versus observation in relapsed chronic lymphocytic leukaemia (PROLONG): an open-label, multicentre, randomised phase 3 study. *Lancet Oncol* 2015;**16**(13): 1370–9.
63. Hillmen P, et al. Chlorambucil plus ofatumumab versus chlorambucil alone in previously untreated patients with chronic lymphocytic leukaemia (COMPLEMENT 1): a randomised, multicentre, open-label phase 3 trial. *Lancet* 2015;**385**(9980):1873–83.
64. Wierda WG, et al. Ofatumumab as single-agent CD20 immunotherapy in fludarabine-refractory chronic lymphocytic leukemia. *J Clin Oncol* 2010;**28**(10):1749–55.
65. Van Cutsem E, et al. Open-Label phase III trial of panitumumab plus Best supportive care compared with Best supportive care alone in patients with chemotherapy-refractory metastatic colorectal cancer. *J Clin Oncol* 2007;**25**(13):1658–64.
66. Van Cutsem E, et al. Randomized phase III study of irinotecan and 5-FU/FA with or without cetuximab in the first-line treatment of patients with metastatic colorectal cancer (mCRC): the CRYSTAL trial. *J Clin Oncol* 2007;**25**(18 Suppl.):4000.
67. Vermorken JB, et al. Platinum-Based chemotherapy plus cetuximab in head and neck cancer. *N Engl J Med* 2008;**359**(11):1116–27.
68. Bonner JA, et al. Radiotherapy plus cetuximab for squamous-cell carcinoma of the head and neck. *N Engl J Med* 2006;**354**(6):567–78.
69. Cunningham D, et al. Cetuximab monotherapy and cetuximab plus irinotecan in irinotecan-refractory metastatic colorectal cancer. *N Engl J Med* 2004;**351**(4):337–45.
70. Roozeboom MH, et al. Overall treatment success after treatment of primary superficial basal cell carcinoma: a systematic review and meta-analysis of randomized and nonrandomized trials. *Br J Dermatol* 2012; **167**(4):733–56.
71. Pujade-Lauraine E, et al. Bevacizumab combined with chemotherapy for platinum-resistant recurrent ovarian cancer: the AURELIA open-label randomized phase III trial. *J Clin Oncol* 2014;**32**(13):1302–8.
72. Tewari KS, et al. Incorporation of bevacizumab in the treatment of recurrent and metastatic cervical cancer: a phase III randomized trial of the Gynecologic Oncology Group. *J Clin Oncol* 2013;**31**(18 Suppl.):3.
73. Bennouna J, et al. Continuation of bevacizumab after first progression in metastatic colorectal cancer (ML18147): a randomised phase 3 trial. *Lancet Oncol* 2013;**14**(1):29–37.
74. Escudier B, et al. Bevacizumab plus interferon alfa-2a for treatment of metastatic renal cell carcinoma: a randomised, double-blind phase III trial. *Lancet* 2007;**370**(9605):2103–11.
75. Kreisl TN, et al. Phase II trial of single-agent bevacizumab followed by bevacizumab plus irinotecan at tumor progression in recurrent glioblastoma. *J Clin Oncol* 2009;**27**(5):740–5.
76. Friedman HS, et al. Bevacizumab alone and in combination with irinotecan in recurrent glioblastoma. *J Clin Oncol* 2009;**27**(28):4733–40.
77. Sandler A, et al. Paclitaxel–carboplatin alone or with Bevacizumab for non–small-cell lung cancer. *N Engl J Med* 2006;**355**(24):2542–50.

78. Giantonio BJ, et al. Bevacizumab in combination with oxaliplatin, fluorouracil, and leucovorin (FOLFOX4) for previously treated metastatic colorectal cancer: results from the Eastern cooperative oncology group study E3200. *J Clin Oncol* 2007;**25**(12):1539–44.
79. Hurwitz H, et al. Bevacizumab plus irinotecan, fluorouracil, and leucovorin for metastatic colorectal cancer. *N Engl J Med* 2004;**350**(23):2335–42.
80. Wahl RL. Tositumomab and (131)I therapy in non-Hodgkin's lymphoma. *J Nucl Med* 2005;**46**(Suppl. 1). 128s–40s.
81. Alcindor T, Witzig TE. Radioimmunotherapy with yttrium-90 ibritumomab tiuxetan for patients with relapsed CD20+ B-cell non-Hodgkin's lymphoma. *Curr Treat Options Oncol* 2002;**3**(4):275–82.
82. Witzig TE, et al. Treatment with ibritumomab tiuxetan radioimmunotherapy in patients with rituximab-refractory follicular non-Hodgkin's lymphoma. *J Clin Oncol* 2002;**20**(15):3262–9.
83. Hillmen P, et al. Alemtuzumab compared with chlorambucil as first-line therapy for chronic lymphocytic leukemia. *J Clin Oncol* 2007;**25**(35):5616–23.
84. Castaigne S, et al. Effect of gemtuzumab ozogamicin on survival of adult patients with de-novo acute myeloid leukaemia (ALFA-0701): a randomised, open-label, phase 3 study. *Lancet* 2012;**379**(9825): 1508–16.
85. Taksin AL, et al. High efficacy and safety profile of fractionated doses of Mylotarg as induction therapy in patients with relapsed acute myeloblastic leukemia: a prospective study of the alfa group. *Leukemia* 2007; **21**(1):66–71.
86. Amadori S, et al. Gemtuzumab ozogamicin versus Best supportive care in older patients with newly diagnosed acute myeloid leukemia unsuitable for intensive chemotherapy: results of the randomized phase III EORTC-GIMEMA AML-19 trial. *J Clin Oncol* 2016;**34**(9):972–9.
87. Verma S, et al. Trastuzumab emtansine for HER2-positive advanced Breast cancer. *N Engl J Med* 2012; **367**(19):1783–91.
88. Bang Y-J, et al. Trastuzumab in combination with chemotherapy versus chemotherapy alone for treatment of HER2-positive advanced gastric or gastro-oesophageal junction cancer (ToGA): a phase 3, open-label, randomised controlled trial. *Lancet* 2010;**376**(9742):687–97.
89. Perez EA, et al. Updated results of the combined analysis of NCCTG N9831 and NSABP B-31 adjuvant chemotherapy with/without trastuzumab in patients with HER2-positive breast cancer. *J Clin Oncol* 2007; **25**(18 Suppl.):512.
90. Goldenberg MM. Trastuzumab, a recombinant DNA-derived humanized monoclonal antibody, a novel agent for the treatment of metastatic breast cancer. *Clin Ther* 1999;**21**(2):309–18.
91. Davies A, et al. Efficacy and safety of subcutaneous rituximab versus intravenous rituximab for first-line treatment of follicular lymphoma (SABRINA): a randomised, open-label, phase 3 trial. *Lancet Haematol* 2017;**4**(6):e272–82.
92. Lugtenburg PJ, et al. Rituximab SC and IV plus CHOP show similar efficacy and safety in the randomised MabEase study in first-line DLBCL. *Hematol Oncol* 2017;**35**:185–6.
93. Salles G, et al. Rituximab maintenance for 2 years in patients with high tumour burden follicular lymphoma responding to rituximab plus chemotherapy (PRIMA): a phase 3, randomised controlled trial. *Lancet* 2011; **377**(9759):42–51.
94. Hallek M, et al. Addition of rituximab to fludarabine and cyclophosphamide in patients with chronic lymphocytic leukaemia: a randomised, open-label, phase 3 trial. *Lancet* 2010;**376**(9747):1164–74.
95. Marcus R, et al. CVP chemotherapy plus rituximab compared with CVP as first-line treatment for advanced follicular lymphoma. *Blood* 2005;**105**(4):1417–23.
96. Feugier P, et al. Long-Term Results of the R-CHOP Study in the Treatment of Elderly Patients With Diffuse Large B-Cell Lymphoma: a Study by the Groupe d'Etude des Lymphomes de l'Adulte. *J Clin Oncol* 2005; **23**(18):4117–26.

97. Pfreundschuh M, et al. CHOP-like chemotherapy plus rituximab versus CHOP-like chemotherapy alone in young patients with good-prognosis diffuse large-B-cell lymphoma: a randomised controlled trial by the MabThera International Trial (MInT) Group. *Lancet Oncol* 2006;**7**(5):379–91.
98. Habermann TM, et al. Rituximab-CHOP versus CHOP alone or with maintenance rituximab in older patients with diffuse large B-cell lymphoma. *J Clin Oncol* 2006;**24**(19):3121–7.
99. Grillo-Lopez AJ, et al. Rituximab: the first monoclonal antibody approved for the treatment of lymphoma. *Curr Pharm Biotechnol* 2000;**1**(1):1–9.
100. Bhatia S, Tykodi SS, Thompson JA. Treatment of metastatic melanoma: an overview. *Oncology* 2009;**23**(6): 488–96.
101. Dutcher JP. Current status of interleukin-2 therapy for metastatic renal cell carcinoma and metastatic melanoma. *Oncology* 2002;**16**(11 Suppl. 13):4–10.
102. Morales A, Nickel JC. Immunotherapy of superficial bladder cancer with BCG. *World J Urol* 1986;**3**(4): 209–14.
103. Kirkwood JM, et al. Interferon alfa-2b adjuvant therapy of high-risk resected cutaneous melanoma: the Eastern Cooperative Oncology Group Trial EST 1684. *J Clin Oncol* 1996;**14**(1):7–17.
104. Quesada JR, et al. Treatment of hairy cell leukemia with recombinant alpha-interferon. *Blood* 1986;**68**(2): 493–7.
105. Papaioannou NE, et al. Harnessing the immune system to improve cancer therapy. *Ann Transl Med* 2016; **4**(14):261.
106. Schuster M, Nechansky A, Kircheis R. Cancer immunotherapy. *Biotechnol J* 2006;**1**(2):138–47.
107. Parker BS, Rautela J, Hertzog PJ. Antitumour actions of interferons: implications for cancer therapy. *Nat Rev Cancer* 2016;**16**:131.
108. Assier E, et al. NK cells and polymorphonuclear neutrophils are both critical for IL-2-induced pulmonary vascular leak syndrome. *J Immunol* 2004;**172**(12):7661–8.
109. Skrombolas D, Frelinger JG. Challenges and developing solutions for increasing the benefits of IL-2 treatment in tumor therapy. *Expert Rev Clin Immunol* 2014;**10**(2):207–17.
110. Jiang T, Zhou C, Ren S. Role of IL-2 in cancer immunotherapy. *Oncoimmunology* 2016;**5**(6):e1163462.
111. Kilinc MO, et al. Reversing tumor immune suppression with intratumoral IL-12: activation of tumor-associated T effector/memory cells, induction of T suppressor apoptosis, and infiltration of CD8+ T effectors. *J Immunol* 2006;**177**(10):6962–73.
112. Eccles SA. Monoclonal antibodies targeting cancer: 'magic bullets' or just the trigger? *Breast Cancer Res* 2000;**3**(2):86.
113. Nimmerjahn F, Ravetch JV. Fcgamma receptors: old friends and new family members. *Immunity* 2006; **24**(1):19–28.
114. Seshacharyulu P, et al. Targeting the EGFR signaling pathway in cancer therapy. *Expert Opin Ther Targets* 2012;**16**(1):15–31.
115. Ferrara N, Adamis AP. Ten years of anti-vascular endothelial growth factor therapy. *Nat Rev Drug Discov* 2016;**15**:385.
116. Janik JE, et al. 90Y-daclizumab, an anti-CD25 monoclonal antibody, provided responses in 50% of patients with relapsed Hodgkin's lymphoma. *Proc Natl Acad Sci U S A* 2015;**112**(42):13045–50.
117. Alderson RF, et al. Characterization of a CC49-based single-chain fragment-beta-lactamase fusion protein for antibody-directed enzyme prodrug therapy (ADEPT). *Bioconjugate Chem* 2006;**17**(2):410–8.
118. Kawashima H. Radioimmunotherapy: a specific treatment protocol for cancer by cytotoxic radioisotopes conjugated to antibodies. *Sci World J* 2014;**2014**:10.
119. Borcoman E, Le Tourneau C. Antibody drug conjugates: the future of chemotherapy? *Curr Opin Oncol* 2016;**28**(5):429–36.

120. Velasquez MP, Bonifant CL, Gottschalk S. Redirecting T cells to hematological malignancies with bispecific antibodies. *Blood* 2017;**131**.
121. Parkhurst MR, et al. Adoptive transfer of autologous natural killer cells leads to high levels of circulating natural killer cells but does not mediate tumor regression. *Clin Cancer Res* 2011;**17**(19):6287–97.
122. Takimoto R, et al. Efficacy of adoptive immune-cell therapy in patients with advanced gastric cancer: a retrospective study. *Anticancer Res* 2017;**37**(7):3947–54.
123. Perica K, Varela JC, Oelke M, Schneck J. Adoptive T cell immunotherapy for cancer. *Rambam Maimonides Med J* 2015;**6**(1):e0004. https://doi.org/10.5041/RMMJ.10179.
124. Rosenberg SA, et al. Durable complete responses in heavily pretreated patients with metastatic melanoma using T cell transfer immunotherapy. *Clin Cancer Res* 2011;**17**(13):4550–7.
125. Wei C, et al. The CIK cells stimulated with combination of IL-2 and IL-15 provide an improved cytotoxic capacity against human lung adenocarcinoma. *Tumour Biol* 2014;**35**(3):1997–2007.
126. West EJ, et al. Immune activation by combination human lymphokine-activated killer and dendritic cell therapy. *Br J Cancer* 2011;**105**:787.
127. Wank R, et al. Benefits of a continuous therapy for cancer patients with a novel adoptive cell therapy by cascade priming (CAPRI). *Immunotherapy* 2014;**6**(3):269–82.
128. Guo Y, Han W. Cytokine-induced killer (CIK) cells: from basic research to clinical translation. *Chin J Cancer* 2015;**34**:6.
129. Laumbacher B, Gu S, Wank R. Activated monocytes prime naïve T cells against autologous cancer: vigorous cancer destruction in vitro and in vivo. *Scand J Immunol* 2012;**75**(3):314–28.
130. Yousefi H, et al. Immunotherapy of cancers comes of age. *Expert Rev Clin Immunol* 2017;**13**(10):1001–15.
131. Chmielewski M, Hombach AA, Abken H. Of CARs and TRUCKs: chimeric antigen receptor (CAR) T cells engineered with an inducible cytokine to modulate the tumor stroma. *Immunol Rev* 2014;**257**(1):83–90.
132. Chmielewski M, et al. IL-12 release by engineered T cells expressing chimeric antigen receptors can effectively Muster an antigen-independent macrophage response on tumor cells that have shut down tumor antigen expression. *Cancer Res* 2011;**71**(17):5697–706.
133. Alatrash G, et al. Cancer immunotherapies, their safety and toxicity. *Expert Opin Drug Saf* 2013;**12**(5): 631–45.
134. Li W, et al. Peptide vaccine: progress and challenges. *Vaccines* 2014;**2**(3):515–36.
135. Kawahara M, Takaku H. A tumor lysate is an effective vaccine antigen for the stimulation of CD4(+) T-cell function and subsequent induction of antitumor immunity mediated by CD8(+) T cells. *Cancer Biol Ther* 2015;**16**(11):1616–25.
136. Srivatsan S, et al. Allogeneic tumor cell vaccines: the promise and limitations in clinical trials. *Hum Vaccines Immunother* 2014;**10**(1):52–63.
137. Nemunaitis J. Vaccines in cancer: GVAX, a GM-CSF gene vaccine. *Expert Rev Vaccines* 2005;**4**(3): 259–74.
138. Palucka K, Banchereau J. Cancer immunotherapy via dendritic cells. Nature reviews. *Cancer* 2012;**12**(4): 265–77.
139. Michiels A, et al. Delivery of tumor-antigen-encoding mRNA into dendritic cells for vaccination. *Methods Mol Biol* 2008;**423**:155–63.
140. Pecher G, et al. Mucin gene (MUC1) transfected dendritic cells as vaccine: results of a phase I/II clinical trial. *Cancer Immunol Immunother* 2002;**51**(11–12):669–73.
141. Butterfield LH, et al. Adenovirus MART-1-engineered autologous dendritic cell vaccine for metastatic melanoma. *J Immunother* 2008;**31**(3):294–309.
142. Matsushita H, et al. A pilot study of autologous tumor lysate-loaded dendritic cell vaccination combined with sunitinib for metastatic renal cell carcinoma. *J Immunother Cancer* 2014;**2**:30.

143. Datta J, et al. Optimizing dendritic cell-Based approaches for cancer immunotherapy. *Yale J Biol Med* 2014; **87**(4):491–518.
144. Aurisicchio L, Ciliberto G. Genetic cancer vaccines: current status and perspectives. *Expert Opin Biol Ther* 2012;**12**(8):1043–58.
145. Tiptiri-Kourpeti A, et al. DNA vaccines to attack cancer: strategies for improving immunogenicity and efficacy. *Pharmacol Ther* 2016;**165**:32–49.
146. Marshall JL, et al. Phase I study in advanced cancer patients of a diversified prime-and-boost vaccination protocol using recombinant vaccinia virus and recombinant nonreplicating avipox virus to elicit anti-carcinoembryonic antigen immune responses. *J Clin Oncol* 2000;**18**(23):3964–73.
147. Blank CU, Enk A. Therapeutic use of anti-CTLA-4 antibodies. *Int Immunol* 2015;**27**(1):3–10.
148. Iwai Y, et al. Cancer immunotherapies targeting the PD-1 signaling pathway. *J Biomed Sci* 2017;**24**(1):26.
149. Kaufman HL, Kohlhapp FJ, Zloza A. Oncolytic viruses: a new class of immunotherapy drugs. *Nat Rev Drug Discov* 2015;**14**:642.
150. Todo T. Oncolytic virus therapy using genetically engineered herpes simplex viruses. *Front Biosci* 2008;**13**: 2060–4.
151. Fukuhara H, Ino Y, Todo T. Oncolytic virus therapy: a new era of cancer treatment at dawn. *Cancer Sci* 2016;**107**(10):1373–9.
152. Emens LA, Middleton G. The interplay of immunotherapy and chemotherapy: harnessing potential synergies. *Cancer Immunol Res* 2015;**3**(5):436–43.
153. Park B, Yee C, Lee K-M. The effect of radiation on the immune response to cancers. *Int J Mol Sci* 2014; **15**(1):927–43.
154. Ott PA, et al. Combination immunotherapy: a road map. *J Immunother Cancer* 2017;**5**(1):16.
155. Hong I-S. Stimulatory versus suppressive effects of GM-CSF on tumor progression in multiple cancer types. *Exp Mol Med* 2016;**48**(7):e242.
156. Lu H. TLR agonists for cancer immunotherapy: tipping the balance between the immune stimulatory and inhibitory effects. *Front Immunol* 2014;**5**(83).
157. Shi M, et al. Application potential of toll-like receptors in cancer immunotherapy: systematic review. *Medicine* 2016;**95**(25):e3951.
158. Byrne KT, Vonderheide RH. CD40 stimulation obviates innate sensors and drives T cell immunity in cancer. *Cell Reports* 2016;**15**(12):2719–32.
159. Vonderheide RH, Glennie MJ. Agonistic CD40 antibodies and cancer therapy. *Clin Cancer Res* 2013;**19**(5): 1035–43.
160. Bevan MJ. Cross-priming. *Nat Immunol* 2006;**7**:363.
161. Seledtsov VI, Goncharov AG, Seledtsova GV. Clinically feasible approaches to potentiating cancer cell-based immunotherapies. *Hum Vaccines Immunother* 2015;**11**(4):851–69.
162. Postow MA, Callahan MK, Wolchok JD. Immune checkpoint Blockade in cancer therapy. *J Clin Oncol* 2015;**33**(17):1974–82.
163. Makkouk A, Chester C, Kohrt HE. Rationale for anti-CD137 cancer immunotherapy. *Eur J Cancer* 2016; **54**(Suppl. C):112–9.
164. Curti BD, et al. OX40 is a potent immune-stimulating target in late-stage cancer patients. *Cancer Res* 2013; **73**(24):7189–98.
165. Thomas LJ, et al. Targeting human CD27 with an agonist antibody stimulates T-cell activation and anti-tumor immunity. *Oncoimmunology* 2014;**3**:e27255.
166. Cooper D, Bansal-Pakala P, Croft M. 4–1BB (CD137) controls the clonal expansion and survival of CD8 T cells in vivo but does not contribute to the development of cytotoxicity. *Eur J Immunol* 2002;**32**(2): 521–9.
167. Rosenberg SA. IL-2: the first effective immunotherapy for human cancer. *J Immunol* 2014;**192**(12):5451–8.

168. Lee S, Margolin K. Cytokines in cancer immunotherapy. *Cancer* 2011;**3**(4):3856−93.
169. Tugues S, et al. New insights into IL-12-mediated tumor suppression. *Cell Death Differ* 2015;**22**(2): 237−46.
170. Lasek W, Zagożdżon R, Jakobisiak M. Interleukin 12: still a promising candidate for tumor immunotherapy? Cancer Immunology. *Immunotherapy* 2014;**63**(5):419−35.
171. Ding Q, et al. CXCL9: evidence and contradictions for its role in tumor progression. *Cancer Med* 2016; **5**(11):3246−59.
172. Mukaida N, Sasaki S-i, Baba T. Chemokines in cancer development and progression and their potential as targeting molecules for cancer treatment. *Mediat Inflamm* 2014;**2014**:170381.
173. Schoenfeld JD, Dranoff G. Anti-angiogenic immunotherapy. *Hum Vaccine* 2011;**7**(9):976−81.
174. Ott PA, Hodi FS, Buchbinder EI. Inhibition of immune checkpoints and vascular endothelial growth factor as combination therapy for metastatic melanoma: an overview of rationale, preclinical evidence, and initial clinical data. *Front Oncol* 2015;**5**:202.
175. Lohmueller J, Finn OJ. Current modalities in cancer immunotherapy: immunomodulatory antibodies, CARs and vaccines. *Pharmacol Therapeut* 2017;**178**(Suppl. C):31−47.
176. Almåsbak H, Aarvak T, Vemuri MC. CAR T cell therapy: a game changer in cancer treatment. *J Immunol Res* 2016;**2016**:5474602.
177. Yoon W, et al. Application of genetically engineered *Salmonella typhimurium* for interferon-gamma-induced therapy against melanoma. *Eur J Cancer* 2017;**70**:48−61.
178. Pardoll DM. The blockade of immune checkpoints in cancer immunotherapy. *Nat Rev Cancer* 2012:12.
179. He J, et al. Development of PD-1/PD-L1 pathway in tumor immune microenvironment and treatment for non-small cell lung cancer. *Sci Rep* 2015;**5**:13110.
180. Alsaab HO, et al. PD-1 and PD-L1 checkpoint signaling inhibition for cancer immunotherapy: mechanism, combinations, and clinical outcome. *Front Pharmacol* 2017;**8**:561.
181. Baitsch L, et al. Exhaustion of tumor-specific CD8(+) T cells in metastases from melanoma patients. *J Clin Invest* 2011;**121**(6):2350−60.
182. Fourcade J, et al. Upregulation of Tim-3 and PD-1 expression is associated with tumor antigen−specific CD8(+) T cell dysfunction in melanoma patients. *J Exp Med* 2010;**207**(10):2175−86.
183. Munn DH, et al. Expression of indoleamine 2,3-dioxygenase by plasmacytoid dendritic cells in tumor-draining lymph nodes. *J Clin Invest* 2004;**114**(2):280−90.
184. Uyttenhove C, et al. Evidence for a tumoral immune resistance mechanism based on tryptophan degradation by indoleamine 2,3-dioxygenase. *Nat Med* 2003;**9**(10):1269−74.
185. Munn DH, et al. Inhibition of T cell proliferation by macrophage tryptophan catabolism. *J Exp Med* 1999; **189**(9):1363−72.
186. Fallarino F, et al. T cell apoptosis by tryptophan catabolism. *Cell Death Differ* 2002;**9**(10):1069−77.
187. Frumento G, et al. Tryptophan-derived catabolites are responsible for inhibition of T and natural killer cell proliferation induced by indoleamine 2,3-dioxygenase. *J Exp Med* 2002;**196**(4):459−68.
188. Prendergast GC, et al. Indoleamine 2,3-dioxygenase pathways of pathgenic inflammation and immune escape in cancer. *Cancer Immunol Immunother* 2014;**63**(7):721−35.
189. Bahary N, et al. Phase 2 trial of the indoleamine 2,3-dioxygenase pathway (IDO) inhibitor indoximod plus gemcitabine/nab-paclitaxel for the treatment of metastatic pancreas cancer: interim analysis. *J Clin Oncol* 2016;**34**(15 Suppl.):3020.
190. Bjoern J, et al. Safety, immune and clinical responses in metastatic melanoma patients vaccinated with a long peptide derived from indoleamine 2,3-dioxygenase in combination with ipilimumab. *Cytotherapy* 2016;**18**(8):1043−55.

CHAPTER

VACCINES, ADJUVANTS, AND DELIVERY SYSTEMS

3

Mahsa Keshavarz-Fathi[1,2,3], **Nima Rezaei**[3,4,5]

School of Medicine, Tehran University of Medical Sciences, Tehran, Iran[1]*; Cancer Immunology Project (CIP), Universal Scientific Education and Research Network (USERN), Tehran, Iran*[2]*; Research Center for Immunodeficiencies, Children's Medical Center, Tehran University of Medical Sciences, Tehran, Iran*[3]*; Department of Immunology, School of Medicine, Tehran University of Medical Sciences, Tehran, Iran*[4]*; Network of Immunity in Infection, Malignancy and Autoimmunity (NIIMA), Universal Scientific Education and Research Network (USERN), Tehran, Iran*[5]

DEFINITION AND CLASSIFICATION

Vaccines are biological products that form an immune protection against an external or internal threat through stimulation of innate and adaptive immunity. Stimulation of the immune response and memory formation are the main fundamentals of vaccination. Vaccines fulfill their goal by prompting the innate immune response and activating antigen presenting cells (APCs), and then causing an adaptive immune response.[1]

Both the innate and adaptive immune systems participate in forming an efficient immune response to an immunization such as vaccine. It is also necessary to induce long-lived stimulation of both the humoral and cell-mediated arms of the adaptive immune system, which is achieved by the production of effector and memory cells of the immune system.[2,3]

Indeed, vaccination is a type of immunization, which itself is divided into two types; i.e., active or passive. Passive immunization is accomplished by administration of preformed antibodies to an unimmunized person. A temporary immunity to a specific threat will be developed following the presence of antibodies in the body and until the time of antibodies' existence. Active immunization develops through the administration of an unimmunized person to a pathogenic component. Thereby, the immune system of the person will be stimulated and provide a long-lived immunity. Both types of the immunization can occur, either naturally, i.e., through natural exposure to the pathogen, for example, available pathogens in the environment, or artificially through medical interventions.[2]

The goal of vaccination is to generate a strong immune response providing long-term protection against infection. In contrast to live pathogens, which are attenuated for vaccination, administration of nonliving vaccines such as inactivated organism or subunit vaccines usually necessitate the addition of an adjuvant to increase the efficacy of vaccine.[4]

As mentioned earlier, nonliving vaccines include inactivated whole pathogen, their toxoid, or only the components of the pathogen, called subunit vaccines. A majority of the vaccines have been prepared by subunit approach. However, some new approaches have been developed, such as DNA

Vaccines for Cancer Immunotherapy. https://doi.org/10.1016/B978-0-12-814039-0.00003-5

vaccines, recombinant vector vaccines, and immune cell vaccines. Selection of an appropriate subunit to induce robust immune response is challenging. To apply a time-consuming approach, whole genome sequence techniques have been used in designing vaccines.[5]

CANCER VACCINE

Vaccination to prevent infectious diseases was a breakthrough to improve public health and increase survival. After prevention of infectious diseases, active involvement of the immune system in the treatment of established diseases such as tumors and infections, i.e., therapeutic vaccine, was put forward. This new concept was more challenging. In the state of an established disease, an immune tolerance or at least a polarized immunity, commonly is present, which certainly influences the direction of the immunity-disease interactions.[6]

The notion of therapeutic vaccines intends to provide not only induction of an immune response but also improvement of an ongoing immune response that is not fully effective. Moreover, alteration of one type of immune response to the desired type can be accomplished through designing therapeutic vaccination. Over the years, many efforts have been made to fulfill the ambition of the therapeutic vaccines. They have the potential to be used in a range of diseases such as infectious diseases and cancers. Vaccines have been used to induce immune tolerance as well. This gives the vaccines the potential to be used for not only infectious diseases but also for any diseases, in which the immune system is involved, such autoimmune diseases and graft rejection.[7–9]

There are antigens on the tumor cells, which are potential targets of the natural immune attack as well as medical interventions, i.e., specific immunotherapy. Selection of an appropriate tumor antigen is of value to push forward an efficient immune response specific for a tumor.[10]

Prevention of some viral diseases also falls in the category of cancer vaccines. Due to the oncogenic properties of some viruses, such as human papillomavirus (HPV), which shows causal etiology of cancers of the cervix uteri, vulva, vagina, penis, anus, and head and neck, prophylactic or preventive vaccines for cancer have been developed.[11] Nonviral cancers are also candidates for preventive vaccine administration.[12] We will discuss the preventive and therapeutic cancer vaccines in the following sections.

PREVENTIVE CANCER VACCINE

It is clear that viral tumors will be prevented by administration of prophylactic vaccines against the carcinogenic viruses. HPV and HBV are the famous examples for this kind of vaccine. Chronic HBV may result in hepatocellular carcinoma (HCC), hence HBV vaccination will reduce the prevalence of HCC. Long-term effects of such immunization have been evaluated in terms of HCC prevention. There was a significant reduction of HCC ($P < .0001$) in vaccinated compared to nonvaccinated infants in Taiwan.[13]

There is a similar success story for HPV vaccine. Gardasil and Cervarix are the two approved vaccines against HPV infection, which demonstrated effective prevention against infection with a number of HPV types, especially the most aggressive and oncogenic types of HPV, i.e., type 16 and type 18. These types are involved in about 65% of cervical cancers. The vaccines prevent HPV infection as well as cervical intraepithelial neoplasia (CIN), adenocarcinoma in situ, and cervical cancer in the countries with good coverage. However, unfortunately it has not been introduced in many parts of the world with very high need.[6,14]

Administration of HPV vaccines started in 2006. It was recommended for 11 to 26-year-old females, and since 2011 it has been recommended for 11 to 21-year-old males as well. After about 10 years, 64% reduction in HPV infection among 14 to 19-year-old females and 34% reduction in 20 to 24-year-old females were reported.[15]

Preventive cancer vaccines are also administered for the premalignant lesions and early stages of cancer. Similar to infection prevention, anti-HPV vaccine is the most studied one for the premalignant and early malignant lesions as well, especially HPV16, responsible for about 50% of cervical carcinomas and 80% of HPV-positive oropharynx cancers.[6]

Vaccines against two proteins of HPV types 16 and 18, i.e., E6 and E7 antigens, play a significant role in the treatment of premalignant lesions, e.g., cervical intraepithelial neoplasia (CIN) or vulvar intraepithelial neoplasia (VIN).[16,17] In one study, peptides from E6 and E7 in addition to incomplete Freund's adjuvant were administered to the women with HPV-16-positive grade 3 VIN. After 24 months, 15 out of 19 patients showed a clinical response and 9 out of 19 were reported as complete responders with clearance of HPV-16. The clinical response showed direct correlation with the T cell responses induced by the vaccine.[18] The efficacy of vaccine against HPV-16 and HPV-18 was tested in the PATRICIA trial on patients with CIN grade 3 and adenocarcinoma in situ of the cervix. This was a randomized double-blind study, which resulted in 100% efficacy of the vaccine in clearance of the premalignant and malignant lesions.[19]

Preventive cancer vaccines also have the potential to be applied for nonviral malignancies. The notion of preventive interventions in individuals susceptible to malignancy is not new and is not limited to the immunotherapy. Chemoprevention such as using tamoxifen, a selective estrogen-receptor modulator, is a famous example.[20]

Since the immunosuppressive state of the tumor microenvironment in the cancer treatment disables the vaccines to function with high efficacy, administration of cancer vaccines at the preventive state might work efficiently.[21]

In order to design preventive vaccines against nonviral cancers, questions have been raised on the selection of an appropriate tumor antigen. An impeccable antigen must not make a cross-reaction with a self-antigen and must not lead to damage to normal tissues. Although the majority of the tumor antigens used as therapeutic vaccines have not resulted in autoimmunity and tissue damage, some differentiation antigens such as melanoma antigens, which cause vitiligo, have gotten the majority of attention.[22] In case of designing a preventive vaccine, this feature must get more attention since there might not be a clinically evident disease and causing severe side effects by prevention is not acceptable.

Advancement of technologies to use new medical modalities have made possible diagnose of premalignant lesions. This assist all types of cancer prevention, including immunoprevention. In order to prevent progress of early cancer or pre-cancerous lesions, either surgical resection or frequent screening are usually performed. However, recurrence of the lesions in those with surgical resection or progress to cancer in those that have not been undergone any prevention is common. Biological methods such as immunoprevention might demonstrate a high capacitance as cancer prevention. Administration of vaccines at the pre-cancerous or early cancer stages capacitates the elimination phase of the immunosurveillance. This could be used either alone or after surgical resection to prevent recurrence of the lesion. There are many benefits for this method, since any contact with the abnormal cells will be decreased, consequently a vigilant immune system will be exhausted later, and the tumor microenvironment will be less immunosuppressive. Altogether, vigorous T cell response will be induce following vaccine administration.[23]

Using preventive vaccines to boost the preexisting immunosurveillance is remarkable even in the case of surgical resection. Tissues adjacent to the early or premalignant lesions are not the same as normal tissues since they usually show alterations in gene expression. Adjacent tissues to colon polyps, which had normal appearance, showed gene alterations similar to the polyps. The genes were responsible for a few functions including immune response. Vaccination could induce a vigorous immune memory response to prevent recurrence of the colon polyps.[23]

THERAPEUTIC CANCER VACCINE

Therapeutic cancer vaccine, for nonviral cancers, targets tumor antigens from various classifications including differentiation antigens, overexpressed antigens, or cancer testis (CT) antigens. The repertoire of T cells, which recognizes these antigens, is formed by central immunological tolerance. Therefore, the majority of high-avidity T cells specific to recognizing and attacking these antigens, which are commonly self-antigens, have been omitted by central tolerance mechanisms.[24,25] T cells that have been induced by therapeutic cancer vaccines must depend on the repertoire created by the central tolerance. Tumor antigens such as overexpressed, CT, or differentiation antigens are all self-antigens, toward which deletional immunological tolerance has been occured as a rule. All of these antigens usually exist in the medullary epithelial cells of the thymus; however, in case of any defects in their expression, an exception for central tolerance will occur.[26–28] Selection of viral- or mutation-based neo-antigens, which are not affected by the central tolerance, leads to designing noteworthy vaccines.

Antigen identification could be bypassed by utilizing a different vaccine type, i.e., administration of autologous whole tumor cell vaccine. In this approach, all of the epitopes of neo-antigens, which are not identified directly in other tumors, are utilized to induce immune response. These neo-antigens are produced by mutations and they are unique for the tumor and the patient.[29–31] However, there are some downsides for this approach, for example, no clear description is provided for the effectiveness of the mentioned epitopes for dendritic cell (DC) presentation, and as the whole tumor cell contains auto-antigens, theoretically, autoimmunity toward these antigens might develop.[6]

In order to design an effective therapeutic cancer vaccine, plenty of tumor antigens must be delivered to DCs to form adequate antigen concentration and proper DC activation.[32] This could result in induction of both CD4+ T cells and T cytotoxic responses. In addition to CD8+ T cells, CD4+ T cells are also required for inducing and maintaining the optimal effector CD8+ T cell responses and CD8+ memory T cells. However, short-term CD8+ responses are induced by selection of an antigen specific for MHC-I and, consequently, the absence of CD4+ T cells.[33–35]

Efficient cancer vaccines could induce both robust effector CD4+ and CD8+ T cells as well as memory responses. Synthetic long peptides and genetic vaccines including RNA, DNA vaccines coadministered with appropriate adjuvants such as toll-like receptors (TLRs) are powerful versions to achieve these responses.[36,37]

In other words, to design an effective cancer vaccine, three main requirements must be met: (1) selection of the appropriate antigen, (2) administration of the proper adjuvant, and (3) development of robust immune responses.

The word *adjuvant* originates from the word *adjuvare*, which denotes "help." Adjuvants are molecules that help the vaccines to induce a strong antigen-specific immune response through activation of antigen presenting cells (APCs). This is yielded through functioning as, or influencing the damage-associated molecular patterns (DAMPs) and/or pathogen-associated molecular patterns (PAMPs).[38,39] These molecules in turn induce activation of pattern recognition receptors of the innate immune system, i.e., TLRs, nucleotide-binding oligomerization domain (NOD)-like receptors (NLRs), absent in melanoma 2 (AIM2)-like receptors (ALRs), C-type lectin receptors (CLRs), or retinoic acid-inducible gene I (RIG-1)-like receptors (RLRs).[40] Adjuvants can also induce cells of the innate immune system such as natural killer (NK) cells to generate cytokines that can improve survival of T cells which are antigen-specific.[39] Adjuvants might work as delivery system as well.[41] There are many examples of adjuvants, some of which are approved and others are being tested in preclinical and clinical studies. In the next section, we will discuss the vaccine adjuvants.

ADJUVANTS

The central objective of administration of a therapeutic cancer vaccine is to activate the antigen-specific T cells, especially the cytotoxic T lymphocytes (CTLs), and to support their proliferation, which eventually leads to destroying the tumor cells and improvement of clinical outcomes. A perfect vaccine adjuvant is expected to perform three actions: (1) provide a satisfactory amount of antigens in the right site and concentration, to be presented by APCs; (2) increase the production of costimulatory molecules; and (3) provide stimulatory cytokines. These three items constitute the three required signals for T cells to recognize the antigen and attack the tumor cells.[42] Indeed, adjuvants decrease the required amount of the antigens or the frequent administration of vaccine, through improving the immunization and increasing the immunogenicity of the antigen.[43] Various categories of adjuvants will be described in the following section.

NONSPECIFIC ADJUVANTS

Nonspecific adjuvants, which could modulate the immune response and act as an antigen delivery system, include mineral salts, emulsions, liposomes, and microparticles.

Aluminum compounds were the first adjuvants discovered in 1926, and have been the main vaccine adjuvants for many years. The next adjuvants, such as Complete Freund's Adjuvant (CFA) and lipopolysaccharide (LPS), were not suitable for humans due to their toxicity profile.[43] Aluminum salts such as $Al(OH)_3$ and $AlPO_4$ or alum are adjuvants, to which antigens are adsorbed and presented to APCs in a multivalent and immunogenic form.[44] Alum induces Th2-type responses and secretion of IL-4, IL-5, IL-6, and IgG1.[52] It also could play a role in activation of proinflammatory cytokines through targeting NOD-like receptor protein 3 (NLRP3) and forming inflammasome.[45,46]

Montanide is a water-in-oil emulsion and the clinical-grade incomplete Freund's adjuvant (IFA). ISA 51 and 720 are a mixture of mineral oil and surfactant mannide monooleate in a 1:1 ratio. They have similarities although the ISA 720 has quick antigen release. Overall, Montanide is similar to IFA; however, it is much less toxic than IFA due to its biodegradable feature.[47] Montanide emulsions form a depot in the injection site, which improves the immune response.[48] Both Montanide ISA 51 and 720 are involved in production of high levels of Ab and CTL responses.[49] MF59 is an oil-in-water emulsion composed of small uniform microvesicles, which induce CD4+ memory T cells and humoral immunity.[49] QS21 is a highly purified immunogenic substance derived from the soap bark tree, which acts as a powerful adjuvant. It is capable of enhancing antigen presentation by APCs and consequently, inducing CTLs. This adjuvant stimulates the production of IL-2, IFN-γ, and IgG2a Abs.[50]

Keyhole limpet hemocyanin (KLH) is a protein originated from the marine mollusk keyhole limpet. Many small molecules including peptides benefit from this protein as a hapten.[49]

CYTOKINES AND CHEMOKINES

Cytokines have been applied as vaccine adjuvants in preclinical and clinical studies to induce the desired immune response. Granulocyte-macrophage colony stimulating factor (GM-CSF), which plays a role in differentiation and proliferation of hematopoietic progenitor cells, can induce maturation of DCs and other APCs. It also empowers APCs for process and presentation of antigens, and consequently, developing robust CTL response against tumor cells.[51,52] Interleukin 12 (IL-12) is one of the cytokines acting as adjuvant through induction of NK, NKT, and T cells to secrete IFN-γ. Following this process, Th1 polarization and CTL response resulted.[53,54] IL-18, a member of IL-1 family, activates macrophages, NKs, and T cells and induces production of IFN-γ.[55]

IL-2, IL-4, IL-7, IL-9, IL-15, and IL-21 belong to the γ-chain family of cytokines. IL-2 is a principal factor for clonal expansion of lymphocytes activated in response to antigens and the induced memory cells. Hemostasis of regulatory T cells (Tregs) is also regulated by IL-2.[56] IL-2 is applied *in vitro* for expansion of T cells, which are infused to the patient as adoptive cell therapy. Nevertheless, *in vivo* application of IL-2 has been restricted due to systemic toxicities.[57] IL-15 acts as a T-cell growth factor as well, but different from IL-2 and through enhancement and homeostasis of memory CD8+ T cells. IL-7 is the next T cell growth factor, which supports the proliferation of naïve T cells. Therefore, it is responsible for enhancement of T cell repertoire. The two later cytokines, IL-7 and IL-15, are able to evert T cell anergy and potentially improve the profile of tumor microenvironment.[53,56] IL-21, which is secreted by CD4+ T cells, stimulates the differentiation of proinflammatory CD4+ Th17 cells and proliferation of memory and naive CD8+ T cell synergistically with IL-7 or IL-15. It is applied for *in vitro* generation of antigen-specific CTLs because it can enhance the amount of expansion and level of affinity of CTLs. IL-21 also has an inhibitory function on the immature DCs.[58,59]

Interferons (IFNs) are classified into two main categories: type I (including α and β) and type II (γ). These glycoproteins are secreted by various cell types and in response to viruses, double-stranded RNAs, and mitogens. IFN-α is produced by macrophages and lymphocytes; IFN-β is secreted from fibroblasts and epithelial cells; and Th1, CTLs, CD8+ NK cells, and lymphokine-activated killer cells are responsible for production of IFN-γ.[56,60] IFNs perform various antitumor activates, which have been reviewed in Chapters 1 and 2.

Chemokines are a useful addition to cancer vaccines because of their role in the recruitment of immune cells to the microenvironment. They function in different settings and exert their inflammatory (inducible) or homeostatic (constitutive) roles. In the inflammatory microenvironment, the inflammatory chemokines are released by the resident or infiltrating cells under the effects of proinflammatory cytokines. They are responsible for recruiting the effector immune cells to the site of inflammation.[61,62]

STIMULATORS OF THE INNATE IMMUNE SYSTEM

C-type Lectin Receptor Ligands

C-type lectin receptors (CLRs) are responsible for recognizing different carbohydrate motifs on microbes through carbohydrate-binding domains. Dectin-1, Dectin-2, Mincle, and CLEC9A are examples of CLRs. β-glucan, one example of the CLR ligands on the cell wall of microbes, binds Dectin-1 and activates spleen tyrosine kinase (Syk)—NF-κB axis to activate production of proinflammatory cytokines. Activation of complement system and recruitment of tumoricidal granulocytes are the other functions mentioned for this ligand. However, there is not enough strong evidence for the antitumor activity of β-glucan in clinical trials.[63–66]

RIG-like Receptor Ligands

RIG-I, RIG-like receptors (RLRs), and melanoma differentiation-associated gene 5 (MDA5), are cytoplasmic RNA sensors, which cause expression of type I IFNs. They have caspase activation and recruitment domains (CARDs), and through CARDS, they activate the interferon promoter stimulator 1 (IPS1)—inhibition of IκB kinase/tumor necrosis factor receptor-associated factor family member-associated NF-kappa-B activator (TANK)-binding kinase 1 (TBK1) axis.[63] Polyribosinic:polyribocytidic acid (Poly I:C) is recognized through RLRs and induces IPS1- and type 1 IFN-dependent apoptosis as well.[67]

Stimulator of Interferon Gene Ligands

Stimulator of interferon gene (STING) responds to intracellular DNA through activation of type I IFN production. It is an endoplasmic reticulum—resident adaptor molecule, which activates the TBK1—IRF3 axis. It is responsible for executing the adjuvant functions of DNA-based vaccines and leads to cytotoxic activities against the tumor. A number of STING ligands can also act as antitumor agents. However, they have not been tested in clinical trials as an immunotherapeutic agent.[68–70]

Toll like Receptor Ligands

TLRs can recognize both PAMPs, expressed by microbes, and DAMPs, expressed by stressed, damaged, or death cells, and subsequently inaugurate an inflammatory response to destroy the abnormal cells. They vigorously activate the innate immune response and subsequently the adaptive immune response through expression of costimulatory molecules and secretion of proinflammatory cytokines, especially, type I interferons.[71,72] There is also evidence of immunotherapeutic activity of TLRs ligands apart from their vaccine adjuvant function.[73] Among the 15 TLRs thus far discovered, TLRs 1—10 are expressed in human. The TLRs recognizing PAMPs are mainly located on the cell surface, including the TLRs 1, 2, 4, 5, and 6. The TLRs recognizing DAMPs are intracellular, i.e., the TLRs 3, 7, 8, 9, and 10. Ligands binding to TLRs 2, 3, 4, 7/8, and 9 have shown clinical evidence to be used as adjuvant or immunotherapy.[74]

TLR2 Agonists

TLR2 acts while dimerizing with TLR1 or TLR6. In the form of TLR1—TLR2 heterodimer, it recognizes triacylated lipopeptides of mycoplasma or gram-negative bacteria, and in the form of TLR2—TLR6 heterodimer, diacylated lipopeptides of mycoplasma and gram-positive bacteria are recognized.[75] Therefore, cell wall skeleton of Bacillus Calmette-Guerin (BCG) can act as a TLR2 agonist to activate DCs. Macrophage activating lipopeptide-2 (MALP)-2 , which is a synthetic analog of mycoplasma cell wall lipoprotein, is the other TLR2 agonist that acts through TLR2–TLR6 heterodimer.[76,77] Both cancer vaccine adjuvant and immunotherapeutic agents have been attributed to the TLR2 agonists.[78,79]

TLR3 Agonists

TLR3 is the only TLR that uses a MyD88-independent signaling pathway, i.e., IFN-β (TRIF)-dependent pathway. It binds to double-stranded RNA to recognize virus RNA and signals through a toll/IL-1R domain–containing adaptor. TRIF activates the interferon regulatory transcription factor 3 (IRF), to increase the production of type I IFNs, as well as chemokines dependent on IFNs, for example, CCL5 and IL-6. Ligands binding TLR3 induce maturation of DCs and production of IL-12p70.[74,80]

Poly I:C is a synthetic TLR3 agonist that could enable Th1, CTL, and production of antigen-specific antibody. Moreover, it could enable intracellular RNA sensors, such as RIG-I and MDA5, and lead to adverse effects.[81–83] Ampligen or poly(I:C12U) showed efficiency in DC maturation for *in vitro* processes and as an adjuvant in DC vaccines *in vivo*. It could enhance upregulation of MHC, and costimulatory molecules on the surface of DCs, as well as production of IL-12.[84,85]

TLR4 Agonists

TLR4 is the only TLR that could activate both MyD88-dependent and -independent (TRIF-dependent) signaling pathways.[74] TLR4 in collaboration with TLR2 recognizes LPS of gram-negative bacteria, which induces septic shock. Nevertheless, nontoxic components of LPS, such as monophosphoryl lipid A (MPL) and glucopyranosyl lipid A (GLA), are utilized as vaccine adjuvant.[49,86]

Preclinical studies have demonstrated GLA as a powerful adjuvant, which induces production of Th1 cytokines, which ultimately induce CTLs. MPL is the only TLR4 agonist approved as vaccine adjuvant and has shown immunogenic properties without severe side effects.[87,88]

TLR7 and TLR8 Agonists

TLR7 and TLR8 distinguish single-stranded viral RNA. They also recognize imidazoquinolines, resiquimod, and imiquimod, which are analogs of DNA or RNA oligonucleotides. These agonists stimulate DCs and macrophages to produce IFN-α, IL-6, IL-8, and IL-12, which consequently, result in induction of Th1 cell-mediated immunity.[49,89,90] Imiquimod is one of the known TLR7 ligands, which was firstly used for the treatment of genital warts. It has been approved for the therapy of actinic keratosis, a premalignant lesion on the skin. This substance exerts antiviral and antitumoral functions mainly through the TLR7–MyD88–IRF7 signaling pathway. Type I IFNs are also secreted through an IRF7-dependent pathway[91,92] Furthermore, TMX-101 or vesimune, which is a liquid formulation of imiquimod, is being evaluated for immunotherapy of noninvasive bladder cancer.[93]

TLR9 Agonists

TLR9 distinguishes viral and bacterial DNA because they include unmethylated cytosine-phosphateguanine (CpG) motifs.[75] TLR9 agonists include CpG analogs, which represent short oligo-deoxynucleotides (CpG ODNs). There are three classes of CpG ODNs, which induce various cell types to secrete different cytokines. Class A and C CpG ODNs activate pDCs and B cells to secrete IFN-α, while class B CpG ODNs lead to the maturation of DCs, in a setting with low level of IFN-α. However, all three classes of CpGs activate Th1 responses, which potentiate them as cancer vaccine adjuvants.[49,94] CpG ODNs have been either combined with other adjuvants or conjugated with nanoparticles or tumor antigens to resolve their ineffectiveness.[40]

Other Immunomodulatory Adjuvants

Hormones and neuromediators could also be administered along with cancer vaccines since they can regulate the adaptive immune responses. Some of these hormones, such as indoleamine and thymus hormones, improve the age-related immune impairment or immunosenescence. Immunomodulation through intervening in the function of costimulatory or coinhibitory molecules on immune cells or modifying the immunoinhibitory microenvironment of the tumor via blockade of Tregs and myeloid-derived suppressor cells are the other approaches used for improving cancer vaccine efficacy.[95]

Adjuvant Systems

Adjuvant could be combined to achieve more effective immune responses especially in immunocompromised individuals. Adjuvant combinations or adjuvant systems (AS) are being utilized along with different vaccines. For example, AS02 is an adjuvant system used in combination with MAGE-A3 cancer vaccine in patients with melanoma and non-small cell lung carcinoma. AS02 consists of MPL, QS-21, and oil-in-water emulsions.[96,97]

VECTORS AND DELIVERY SYSTEM

Vectors of antigen enhance the signal 1 needed for antigen recognition by T cells. It is performed by various mechanisms including protection of the antigen from degrading enzymes and increasing the antigen presentation, improvement of small antigen uptake by APC through encompassing the antigen and shaping particles similar to the size of pathogens, and delivery of the antigen to the draining lymph nodes, i.e., the site of antigen presentation to T cells. Some vaccine adjuvant can function as delivery system as well, such as IFA and Montanide formulations, alum, and micro/nano particles.[42]

In addition to the adjuvants mentioned in the previous section, there are other types of adjuvants that could deliver the antigens. Virosomes are the viral particles that are used as delivery system. They lack the genetic material and the nucleocapsid of the virus and can be genetically engineered to convey the antigens with or without adjuvants. Influenza virosomes are one of the known examples utilized for cancer vaccination.[61,98] Liposomes, which are microvesicle composed of a bilayer phospholipid, are competent candidate for serving as a vector to carry the antigen or adjuvant. They have been used widely in animal studies. However, they are not clinically consumed due to expensiveness and concerns about its manufacturing. Interestingly, delivery of CpG ODNs via liposomes leads to an effective adjuvant effect, which is not achieved by CpG ODN alone.[99,100]

Immune stimulating complexes (ISCOMs) originated from Quillaia saponins and contain cholesterol and phospholipid. Their diameter is ~40 nm, which enables it to serve as a suitable delivery system conveying the antigen to cells. They are capable of enhancing antigen cross-presentation by DCs and inducing both cell-mediated and long-lived humoral immunity. ISCOMATRIX exerts the adjuvant functions through NLRP3 inflammasome pathway. TNF-α and Il-18 are involved in the adjuvant role of the ISCOMATRIX.[101–103]

Delivery systems are remarkable determinants of the immunogenicity of genetic vaccines. Novel physical delivery systems showed more benefit than chemical agents did.[104,105] Biolistic gene gun is a delivery system taking the advantage of a high-pressured tool to shoot the naked DNA adhered to gold beads. Through this system, DNA directly reaches the Langerhans cells of the skin. It is capable of activation of more CD8+ T cell.[106] Electroporation is a delivery system that greatly improves uptake of DNA by the cells. It functions as an adjuvant as well, inducing inflammation subsequent to tissue damage, which attracts APCs, lymphocytes, and macrophages to the site of injection. This leads to powerful immune responses.[107] Tattooing is an intradermal vaccine delivery system that is in clinical trials in melanoma patients.[108] This method uses an instrument containing fine metal needles that perforate the skin, using constant high-frequency oscillation to deliver the DNA. It is capable of inducing strong immune responses.[109]

ROUTE OF ADMINISTRATION

APCs loaded with antigens migrate via the afferent lymph to the lymph nodes, the site where they can prime T cells. Afterwards, primed and activated T cells must get access to the tumor cells through migration via the efferent lymph, thoracic duct, and blood.[6] Various routes might be considered for vaccine administration. Cancer vaccines must be administered through a route resulting in effective immunization. The best routes of administration must deliver the antigen to DCs efficiently. Intradermal and intralymphatic administration of vaccines have been reported effective.[110] There are some other effective routes such as subcutaneous delivery of long peptide in Montanide and intramuscular injection of genetic vaccines combined with electroporation.[111] Theoretically, intranodal administration of DC-based vaccines leads to higher frequency of injected DCs traveling to the draining lymph nodes. However, no superiority of this approach compared with other routes has been reported.[112]

REFERENCES

1. Pasquale A, et al. Vaccine adjuvants: from 1920 to 2015 and beyond. *Vaccines* 2015;**3**(2):320.
2. Clem AS. Fundamentals of vaccine immunology. *J Global Infect Dis* 2011;**3**(1):73–8.
3. Gourley TS, et al. Generation and maintenance of immunological memory. *Semin Immunol* 2004;**16**(5): 323–33.
4. Petrovsky N, Aguilar JC. Vaccine adjuvants: current state and future trends. *Immunol Cell Biol* 2004;**82**(5): 488–96.
5. Rappuoli R. Reverse vaccinology, a genome-based approach to vaccine development. *Vaccine* 2001;**19**(17): 2688–91.
6. Melief CJM, et al. Therapeutic cancer vaccines. *J Clin Invest* 2015;**125**(9):3401–12.
7. Finn OJ. Cancer vaccines: between the idea and the reality. *Nat Rev Immunol* 2003;**3**(8):630–41.
8. Xiao BG, Huang YM, Link H. Dendritic cell vaccine design: strategies for eliciting peripheral tolerance as therapy of autoimmune diseases. *BioDrugs* 2003;**17**(2):103–11.
9. Young JW, Merad M, Hart DN. Dendritic cells in transplantation and immune-based therapies. *Biol Blood Marrow Transplant* 2007;**13**(1 Suppl. 1):23–32.
10. Melero I, et al. Therapeutic vaccines for cancer: an overview of clinical trials. *Nat Rev Clin Oncol* 2014; **11**(9):509–24.
11. Bosch FX, et al. Comprehensive control of human papillomavirus infections and related diseases. *Vaccine* 2013;**31**(0 8):I1–31.
12. Lollini P-L, et al. The promise of preventive cancer vaccines. *Vaccines* 2015;**3**(2):467–89.

13. Chang M-H, et al. Long-term effects of hepatitis B immunization of infants in preventing liver cancer. *Gastroenterology* 2016;**151**(3):472–80. e1.
14. Harper DM, DeMars LR. HPV vaccines - a review of the first decade. *Gynecol Oncol* 2017;**146**(1): 196–204.
15. Markowitz LE, et al. Prevalence of HPV after introduction of the vaccination program in the United States. *Pediatrics* 2016;**137**(3).
16. van Poelgeest MI, et al. HPV16 synthetic long peptide (HPV16-SLP) vaccination therapy of patients with advanced or recurrent HPV16-induced gynecological carcinoma, a phase II trial. *J Transl Med* 2013;**11**:88.
17. Kenter GG, et al. Vaccination against HPV-16 oncoproteins for vulvar intraepithelial neoplasia. *N Engl J Med* 2009;**361**(19):1838–47.
18. van Poelgeest MI, et al. Vaccination against oncoproteins of HPV16 for noninvasive vulvar/vaginal lesions: lesion clearance is related to the strength of the T-cell response. *Clin Cancer Res* 2016;**22**(10):2342–50.
19. Lehtinen M, et al. Overall efficacy of HPV-16/18 AS04-adjuvanted vaccine against grade 3 or greater cervical intraepithelial neoplasia: 4-year end-of-study analysis of the randomised, double-blind PATRICIA trial. *Lancet Oncol* 2012;**13**(1):89–99.
20. Lollini PL, et al. Vaccines for tumour prevention. *Nat Rev Cancer* 2006;**6**(3):204–16.
21. Gray A, et al. A paradigm shift in therapeutic vaccination of cancer patients: the need to apply therapeutic vaccination strategies in the preventive setting. *Immunol Rev* 2008;**222**:316–27.
22. Pardoll DM. Inducing autoimmune disease to treat cancer. *Proc Natl Acad Sci Unit States Am* 1999;**96**(10): 5340–2.
23. Lian J, et al. Aberrant gene expression profile of unaffected colon mucosa from patients with unifocal colon polyp. *Med Sci Monit* 2015;**21**:3935–40.
24. Abramson J, et al. Aire's partners in the molecular control of immunological tolerance. *Cell* 2010;**140**(1): 123–35.
25. Klein L, et al. Antigen presentation in the thymus for positive selection and central tolerance induction. *Nat Rev Immunol* 2009;**9**(12):833–44.
26. Nitta T, et al. Thymic microenvironments for T-cell repertoire formation. *Adv Immunol* 2008;**99**:59–94.
27. Bos R, et al. Expression of a natural tumor antigen by thymic epithelial cells impairs the tumor-protective CD4+ T-cell repertoire. *Cancer Res* 2005;**65**(14):6443–9.
28. Pinto S, et al. Misinitiation of intrathymic MART-1 transcription and biased TCR usage explain the high frequency of MART-1-specific T cells. *Eur J Immunol* 2014;**44**(9):2811–21.
29. Wu A, et al. In vivo vaccination with tumor cell lysate plus CpG oligodeoxynucleotides eradicates murine glioblastoma. *J Immunother* 2007;**30**(8):789–97.
30. Reardon DA, et al. An update of vaccine therapy and other immunotherapeutic approaches for glioblastoma. *Expert Rev Vaccine* 2013;**12**(6):597–615.
31. Cadena A, et al. Radiation and anti-cancer vaccines: a winning combination. *Vaccines* 2018;**6**(1):9.
32. Ossendorp F, et al. Specific T helper cell requirement for optimal induction of cytotoxic T lymphocytes against major histocompatibility complex class II negative tumors. *J Exp Med* 1998;**187**(5):693–702.
33. Janssen EM, et al. CD4+ T cells are required for secondary expansion and memory in CD8+ T lymphocytes. *Nature* 2003;**421**(6925):852–6.
34. Xiang J, Huang H, Liu Y. A new dynamic model of CD8+ T effector cell responses via CD4+ T helper-antigen-presenting cells. *J Immunol* 2005;**174**(12):7497–505.
35. Kast WM, et al. Cooperation between cytotoxic and helper T lymphocytes in protection against lethal Sendai virus infection. Protection by T cells is MHC-restricted and MHC-regulated; a model for MHC-disease associations. *J Exp Med* 1986;**164**(3):723–38.
36. Khan S, et al. Distinct uptake mechanisms but similar intracellular processing of two different toll-like receptor ligand-peptide conjugates in dendritic cells. *J Biol Chem* 2007;**282**(29):21145–59.

37. van den Boorn JG, Barchet W, Hartmann G. Nucleic acid adjuvants: toward an educated vaccine. *Adv Immunol* 2012;**114**:1–32.
38. Takeuchi O, Akira S. Pattern recognition receptors and inflammation. *Cell* 2010;**140**(6):805–20.
39. McKee AS, Munks MW, Marrack P. How do adjuvants work? Important considerations for new generation adjuvants. *Immunity* 2007;**27**(5):687–90.
40. Temizoz B, Kuroda E, Ishii KJ. Vaccine adjuvants as potential cancer immunotherapeutics. *Int Immunol* 2016;**28**(7):329–38.
41. Saupe A, et al. Immunostimulatory colloidal delivery systems for cancer vaccines. *Expert Opin Drug Deliv* 2006;**3**(3):345–54.
42. Khong H, Overwijk WW. Adjuvants for peptide-based cancer vaccines. *J Immunother Cancer* 2016;**4**(1):56.
43. Aguilar JC, Rodríguez EG. Vaccine adjuvants revisited. *Vaccine* 2007;**25**(19):3752–62.
44. Noe SM, et al. Mechanism of immunopotentiation by aluminum-containing adjuvants elucidated by the relationship between antigen retention at the inoculation site and the immune response. *Vaccine* 2010; **28**(20):3588–94.
45. Franchi L, Nunez G. The Nlrp3 inflammasome is critical for aluminium hydroxide-mediated IL-1beta secretion but dispensable for adjuvant activity. *Eur J Immunol* 2008;**38**(8):2085–9.
46. Eisenbarth SC, et al. Crucial role for the Nalp3 inflammasome in the immunostimulatory properties of aluminium adjuvants. *Nature* 2008;**453**(7198):1122–6.
47. Scalzo AA, et al. Induction of protective cytotoxic T cells to murine cytomegalovirus by using a nonapeptide and a human-compatible adjuvant (Montanide ISA 720). *J Virol* 1995;**69**(2):1306–9.
48. Miles AP, et al. Montanide ISA 720 vaccines: quality control of emulsions, stability of formulated antigens, and comparative immunogenicity of vaccine formulations. *Vaccine* 2005;**23**(19):2530–9.
49. Chiang CL, Kandalaft LE, Coukos G. Adjuvants for enhancing the immunogenicity of whole tumor cell vaccines. *Int Rev Immunol* 2011;**30**(2–3):150–82.
50. Kensil CR, Kammer R. QS-21: a water-soluble triterpene glycoside adjuvant. *Expert Opin Invest Drugs* 1998;**7**(9):1475–82.
51. Jinushi M, Hodi FS, Dranoff G. Enhancing the clinical activity of granulocyte-macrophage colony-stimulating factor-secreting tumor cell vaccines. *Immunol Rev* 2008;**222**:287–98.
52. Dranoff G. GM-CSF-based cancer vaccines. *Immunol Rev* 2002;**188**:147–54.
53. Stevceva L, Moniuszko M, Ferrari MG. Utilizing IL-12, IL-15 and IL-7 as mucosal vaccine adjuvants. *Lett Drug Des Discov* 2006;**3**(8):586–92.
54. Baxevanis CN, Perez SA, Papamichail M. Cancer immunotherapy. *Crit Rev Clin Lab Sci* 2009;**46**(4): 167–89.
55. Srivastava S, Salim N, Robertson MJ. Interleukin-18: biology and role in the immunotherapy of cancer. *Curr Med Chem* 2010;**17**(29):3353–7.
56. Lee S, Margolin K. Cytokines in cancer immunotherapy. *Cancers* 2011;**3**(4):3856–93.
57. Palena C, Schlom J. Vaccines against human carcinomas: strategies to improve antitumor immune responses. *J Biomed Biotechnol* 2010;**2010**:380697.
58. Spolski R, Leonard WJ. The Yin and Yang of interleukin-21 in allergy, autoimmunity and cancer. *Curr Opin Immunol* 2008;**20**(3):295–301.
59. Davis ID, et al. Interleukin-21 signaling: functions in cancer and autoimmunity. *Clin Cancer Res* 2007; **13**(23):6926–32.
60. Bekisz J, et al. Immunomodulatory effects of interferons in malignancies. *J Interferon Cytokine Res* 2013; **33**(4):154–61.
61. Bobanga ID, Petrosiute A, Huang AY. Chemokines as cancer vaccine adjuvants. *Vaccines (Basel)* 2013;**1**(4): 444–62.

62. Mohan T, et al. Applications of chemokines as adjuvants for vaccine immunotherapy. *Immunobiology* 2018; **223**(6–7):477–85.
63. Akira S. Innate immunity and adjuvants. *Philos Trans R Soc Lond B Biol Sci* 2011;**366**(1579):2748–55.
64. Chan GC, Chan WK, Sze DM. The effects of beta-glucan on human immune and cancer cells. *J Hematol Oncol* 2009;**2**:25.
65. Akramiene D, et al. Effects of beta-glucans on the immune system. *Medicina (Kaunas)* 2007;**43**(8): 597–606.
66. Go P, Chung CH. Adjuvant PSK immunotherapy in patients with carcinoma of the nasopharynx. *J Int Med Res* 1989;**17**(2):141–9.
67. Zitvogel L, Kroemer G. Anticancer immunochemotherapy using adjuvants with direct cytotoxic effects. *J Clin Invest* 2009;**119**(8):2127–30.
68. Ishikawa H, Ma Z, Barber GN. STING regulates intracellular DNA-mediated, type I interferon-dependent innate immunity. *Nature* 2009;**461**(7265):788–92.
69. Woo SR, et al. STING-dependent cytosolic DNA sensing mediates innate immune recognition of immunogenic tumors. *Immunity* 2014;**41**(5):830–42.
70. Conlon J, et al. Mouse, but not human STING, binds and signals in response to the vascular disrupting agent 5,6-dimethylxanthenone-4-acetic acid. *J Immunol* 2013;**190**(10):5216–25.
71. Dubensky Jr TW, Reed SG. Adjuvants for cancer vaccines. *Semin Immunol* 2010;**22**(3):155–61.
72. Mogensen TH. Pathogen recognition and inflammatory signaling in innate immune defenses. *Clin Microbiol Rev* 2009;**22**(2):240–73.
73. Kaczanowska S, Joseph AM, Davila E. TLR agonists: our best frenemy in cancer immunotherapy. *J Leukoc Biol* 2013;**93**(6):847–63.
74. Khajeh Alizadeh Attar M, et al. *Basic understanding and therapeutic approaches to target toll-like receptors in cancerous microenvironment and metastasis*. 2017.
75. Kawai T, Akira S. The role of pattern-recognition receptors in innate immunity: update on Toll-like receptors. *Nat Immunol* 2010;**11**(5):373–84.
76. Uehori J, et al. Simultaneous blocking of human Toll-like receptors 2 and 4 suppresses myeloid dendritic cell activation induced by *Mycobacterium bovis* bacillus Calmette-Guerin peptidoglycan. *Infect Immun* 2003;**71**(8):4238–49.
77. Schmidt J, et al. Intratumoural injection of the toll-like receptor-2/6 agonist 'macrophage-activating lipopeptide-2' in patients with pancreatic carcinoma: a phase I/II trial. *Br J Cancer* 2007;**97**(5):598–604.
78. Asprodites N, et al. Engagement of Toll-like receptor-2 on cytotoxic T-lymphocytes occurs in vivo and augments antitumor activity. *FASEB J* 2008;**22**(10):3628–37.
79. Ingale S, et al. Robust immune responses elicited by a fully synthetic three-component vaccine. *Nat Chem Biol* 2007;**3**(10):663–7.
80. Verdijk RM, et al. Polyriboinosinic polyribocytidylic acid (poly(I: C)) induces stable maturation of functionally active human dendritic cells. *J Immunol* 1999;**163**(1):57–61.
81. Stahl-Hennig C, et al. Synthetic double-stranded RNAs are adjuvants for the induction of T helper 1 and humoral immune responses to human papillomavirus in rhesus macaques. *PLoS Pathog* 2009;**5**(4): e1000373.
82. Hafner AM, Corthesy B, Merkle HP. Particulate formulations for the delivery of poly(I: C) as vaccine adjuvant. *Adv Drug Deliv Rev* 2013;**65**(10):1386–99.
83. Kato H, et al. Differential roles of MDA5 and RIG-I helicases in the recognition of RNA viruses. *Nature* 2006;**441**(7089):101–5.
84. Jasani B, Navabi H, Adams M. Ampligen: a potential toll-like 3 receptor adjuvant for immunotherapy of cancer. *Vaccine* 2009;**27**(25–26):3401–4.

85. Navabi H, et al. A clinical grade poly I: C-analogue (Ampligen) promotes optimal DC maturation and Th1-type T cell responses of healthy donors and cancer patients in vitro. *Vaccine* 2009;**27**(1):107–15.
86. Fox CB, et al. Synthetic and natural TLR4 agonists as safe and effective vaccine adjuvants. *Subcell Biochem* 2010;**53**:303–21.
87. Coler RN, et al. Development and characterization of synthetic glucopyranosyl lipid adjuvant system as a vaccine adjuvant. *PLoS One* 2011;**6**(1):e16333.
88. Cluff CW. Monophosphoryl lipid A (MPL) as an adjuvant for anti-cancer vaccines: clinical results. *Adv Exp Med Biol* 2010;**667**:111–23.
89. Shi C, et al. Discovery of imidazoquinolines with toll-like receptor 7/8 independent cytokine induction. *ACS Med Chem Lett* 2012;**3**(6):501–4.
90. Agrawal S, Kandimalla ER. Synthetic agonists of Toll-like receptors 7, 8 and 9. *Biochem Soc Trans* 2007;**35**(Pt 6):1461–7.
91. Torres A, et al. Immune-mediated changes in actinic keratosis following topical treatment with imiquimod 5% cream. *J Transl Med* 2007;**5**:7.
92. Petes C, Odoardi N, Gee K. The toll for trafficking: toll-like receptor 7 delivery to the endosome. *Front Immunol* 2017;**8**:1075.
93. Junt T, Barchet W. Translating nucleic acid-sensing pathways into therapies. *Nat Rev Immunol* 2015;**15**(9):529–44.
94. Krieg AM. Development of TLR9 agonists for cancer therapy. *J Clin Invest* 2007;**117**(5):1184–94.
95. Seledtsov VI, Goncharov AG, Seledtsova GV. Clinically feasible approaches to potentiating cancer cell-based immunotherapies. *Hum Vaccines Immunother* 2015;**11**(4):851–69.
96. Garcon N, Van Mechelen M. Recent clinical experience with vaccines using MPL- and QS-21-containing adjuvant systems. *Expert Rev Vaccines* 2011;**10**(4):471–86.
97. Vantomme V, et al. Immunologic analysis of a phase I/II study of vaccination with MAGE-3 protein combined with the AS02B adjuvant in patients with MAGE-3-positive tumors. *J Immunother* 2004;**27**(2):124–35.
98. Daemen T, et al. Virosomes for antigen and DNA delivery. *Adv Drug Deliv Rev* 2005;**57**(3):451–63.
99. Sercombe L, et al. Advances and challenges of liposome assisted drug delivery. *Front Pharmacol* 2015;**6**:286.
100. Brignole C, et al. Therapeutic targeting of TLR9 inhibits cell growth and induces apoptosis in neuroblastoma. *Cancer Res* 2010;**70**(23):9816–26.
101. Cox JC, Sjolander A, Barr IG. ISCOMs and other saponin based adjuvants. *Adv Drug Deliv Rev* 1998;**32**(3):247–71.
102. Wilson NS, et al. Inflammasome-dependent and -independent IL-18 production mediates immunity to the ISCOMATRIX adjuvant. *J Immunol* 2014;**192**(7):3259–68.
103. Sanders MT, et al. ISCOM-based vaccines: the second decade. *Immunol Cell Biol* 2005;**83**(2):119–28.
104. Egan MA, et al. Rational design of a plasmid DNA vaccine capable of eliciting cell-mediated immune responses to multiple HIV antigens in mice. *Vaccine* 2006;**24**(21):4510–23.
105. Drape RJ, et al. Epidermal DNA vaccine for influenza is immunogenic in humans. *Vaccine* 2006;**24**(21):4475–81.
106. Trimble C, et al. Comparison of the CD8+ T cell responses and antitumor effects generated by DNA vaccine administered through gene gun, biojector, and syringe. *Vaccine* 2003;**21**(25–26):4036–42.
107. Liu J, et al. Recruitment of antigen presenting cells to the site of inoculation and augmentation of human immunodeficiency virus type 1 DNA vaccine immunogenicity by in vivo electroporation. *J Virol* 2008;**82**(11):5643–9.
108. Quaak SG, et al. GMP production of pDERMATT for vaccination against melanoma in a phase I clinical trial. *Eur J Pharm Biopharm* 2008;**70**(2):429–38.

109. Bins AD, et al. A rapid and potent DNA vaccination strategy defined by in vivo monitoring of antigen expression. *Nat Med* 2005;**11**(8):899–904.
110. Fong L, et al. Dendritic cells injected via different routes induce immunity in cancer patients. *J Immunol* 2001;**166**(6):4254–9.
111. Schijns V, et al. Immune adjuvants as critical guides directing immunity triggered by therapeutic cancer vaccines. *Cytotherapy* 2014;**16**(4):427–39.
112. Lesterhuis WJ, et al. Route of administration modulates the induction of dendritic cell vaccine–induced antigen-specific T cells in advanced melanoma patients. *Clin Cancer Res* 2011;**17**(17):5725–35.

CHAPTER

TUMOR ANTIGENS

4

Saeed Farajzadeh Valilou[1], **Nima Rezaei**[2,3,4]

Cancer Immunology Project (CIP), Universal Scientific Education and Research Network (USERN), Tehran, Iran[1]; *Research Center for Immunodeficiencies, Children's Medical Center, Tehran University of Medical Sciences, Tehran, Iran*[2]; *Department of Immunology, School of Medicine, Tehran University of Medical Sciences, Tehran, Iran*[3]; *Network of Immunity in Infection, Malignancy and Autoimmunity (NIIMA), Universal Scientific Education and Research Network (USERN), Tehran, Iran*[4]

INTRODUCTION

Cancer vaccination includes a wide range of approaches to generate, amplify, or skew (or a combination of them) antitumor immunity. Administration of tumor antigens, often accompanied with antigen presenting cells (APCs) or other immune modulators, or direct modulation of the tumor, is required to achieve this goal. Importantly, standard approaches for cancer treatments such as chemotherapy, radiotherapy, and surgery, as well as impeding signaling pathways of small molecules via inhibitors, can have similar consequences to cancer vaccines. This is done by causing the release of antigens from dying tumor cells or enhancing the expression of tumor antigens within the tumor and by elevating antitumor immunity for therapeutic benefit.[1,2]

Lurquin et al. revealed the molecular nature of the antigens recognized by cytotoxic T lymphocyte (CTL) on tumors by showing that a mouse tumor-specific CTL recognized a peptide that mutated in cancer cells and derived from a self-protein.[3] This observation showed that major histocompatibility (MHC) class I molecules continuously bound peptides of 8–10 amino acids on the cell surface that are derived from a wide variety of, if not all, intracellular proteins.[4] Some of these peptides stem from altered or aberrantly expressed proteins in tumors, therefore prepare the cells for CTL recognition.[5]

In addition, a few scientists, especially Gross, Foley, Prehn, and Old, reported in the 20th century that tumors, which are nonviral include unique, tumor-specific antigens (TSAs).[6–9] The investigations revealed that when inbred mice with carcinogen-induced tumors undergone surgical resection and were cured, they were immune to next rechallenge with the same tumor cells, but this was not happened with other types of tumor cells, even the tumors induced in the same way in other hosts. Later in the next quarter of the 20th century, the functions of the MHC proteins in the pivotal step of immunity against cancer, i.e. the antigen presentation were figured out,[10,11] methods were developed to propagate antigen-specific CTL in culture,[12,13] and it became feasible to use molecular biology techniques to clone and express gene products. Together, these improvements led to the identification of molecular pattern of TSAs.[14]

Vaccines for Cancer Immunotherapy. https://doi.org/10.1016/B978-0-12-814039-0.00004-7

It is now evident that cytotoxic response of immune system is resulted from the activity of the endogenous T cells, which is based on recognizing cancer cells, in some situations, can control the disease. Preclinical data from numerous mouse models have revealed the critical role of T cells in tumor control via depleting either the adaptive immune response or T cells alone, leading to removal of various degrees of tumor rejection. The importance of T cells in the human setting is demonstrated by the association of tumor-infiltrating T cells with good prognosis in a considerable number of different cancers.[15–18] Galon and colleagues have demonstrated that as the currently used staging system, quantifying the infiltration of antigen experienced CD8+ T cells in the tumor invasive margin, and the center of the tumor applied as a strong prognostic tool.[19,20] Furthermore, Ribas and colleagues showed that tumor infiltration and location of CD8+ cells in human melanoma can function as a predictive biomarker for clinical outcome to anti-PD-1 therapy.[21]

In this chapter the tumor antigens from different aspects, including targeting tumor antigens, epitope spreading, tumor-associated antigens, tumor-specific antigens, antigens with nature of glycolipid and glycoprotein, and mono-epitope versus poly-epitope antigens, will be reviewed comprehensively.

IDENTIFICATION OF TUMOR ANTIGENS

Initially, first strategies that were used to identify tumor antigens were accomplished by screening expression libraries via *in vitro* sensitization of peripheral blood mononuclear cell (PBMCs) or by *in vitro* culture with tumor-infiltrating lymphocytes (TILs) against autologous tumor cells or autologous normal cells that were either transfected with genetic constructs encoding candidate antigens or pulsed with candidate T cell epitopes. The antigens identified using these approaches can be generally classified into five categories: (1) antigens originated from gene products that are widely expressed in normal tissues at relatively low levels in comparison with malignant cells, (2) differentiation antigens expressed at relatively high levels in a single tissue, (3) antigens that are limited in their expression in adults to germ cells that lack MHC expression (cancer germline (CG) or cancer testes antigens), (4) viral antigens, and (5) mutated antigens.[22]

From the samples of patients with cancer, i.e. their blood or tumoral tissue, antitumor CTL clones have been extracted.[23,24] One of the methods usually utilized is expression cloning to find the peptides recognized by such CTL. This approach includes isolating and extracting the peptide-encoding gene by transfecting a library of tumoral cDNA and evaluation of the transfected cells to check if they are capable of activating the CTL clone.[25,26] In order to find the region accounting for the antigenic peptide, fragments of the identified gene can then be transfected, and eventually candidate peptides possessing satisfactory amount of HLA-binding motifs are subjects of evaluation for the capacity of sensitization of cancerous cells to lysis by the CTL. This strategy was successfully applied to detect many antigenic peptides.[25,27–29]

Nowadays, "reverse immunology" strategy is common for finding the appropriate tumor-associated antigenic (TAA) peptides,[30] which includes selection of peptides with enough HLA binding motifs inside a protein of interest, for instance, overexpressed proteins or mutated oncogenes. The peptides with highest ability of binding to HLA are utilized for pulsing APCs, stimulating T lymphocytes *in vitro*, to initiate CTL response specific for the antigen used for stimulation. There is an disadvantage for the explained approach. Tumors might not effectively process the peptides detected through this way. Therefore, verification of recognizing tumor cells, which naturally present the peptide, by CTL is an essential step of this strategy. Moreover, evaluation of transfectants expressing the gene at normal levels[31] or cells with altered expression of the gene applying small interfering (si) or short hairpin (sh) RNAs is necessary.[32]

Elution of antigens from the MHC-I on the cancerous cells is another method to find tumor antigens.[33–36] To find or to assert the relevance of peptides, modified post-translationally mass spectrometry is used and is beneficial.[37] Tandem mass spectrometry with chromatography separation was applied for identification and sequencing of an MHC-bound peptide, i.e. gp100 (280-288) on a melanoma cell line line recognized by five CTL lines.[35] A number of predominant epitopes from gp100 and MART-1was identified through this method.[38] A lot of novel TAAs of renal cell carcinoma have been identified by using peptide elution and mass spectrometry.[39]

Also, expression cloning was used for TAAs recognized by B-cells rather than to identify tumor antigens that triggered T-cell responses.[40] A cancer germline antigen, NY-ESO-1, was identified via this approach.[41] The library was prepared using total RNA extracted from cell lines, malignant and non-malignant tissues of a patient with esophageal squamous cell carcinoma. Building the cDNA library was the next step. The library was expressed in Escherichia coli lysate and assessed for antibody binding from the patient's own serum. Eight genes, i.e. NY-ESO-1-8, were associated with the reactive clones. Reverse transcription polymerase chain reaction (RT-PCR) showed that among the eight gene, NY-ESO-1 is not expressed by normal tissue, but is expressed by the testis and the tumor. About 80% of synovial cell sarcomas as well as 10%–50% of metastatic melanomas, lung, ovarian, and breast cancers have been reported that express NY-ESO-1.[42–44] It is shown that response of high-affinity IgG serum suggests that in these patients, there are at least CD4+ T-cell responses, and subsequently both Class I and Class II responses were delineated and cloned.[45]

Generally, approaches to identify tumor antigens include: (1) reverse immunology by prediction of epitope according to known HLA-binding motifs done by dedicated software and sometimes supported by proteasome-cleavage programs; (2) biochemical methods, which remove and fractionate naturally expressed TAA peptides on tumor cells in the context of HLA molecules by chromatography and mass spectrometry, and (3) technology of DNA microarray, which makes feasible the gene expression profiles to be compared in tumor tissues and normal counterparts and, in this regard, differential display, serial analysis of gene expression, representational difference analysis, and suppression subtractive hybridization could be used.[25,46]

OVEREXPRESSED PROTEINS AND MUTATED ANTIGENS IN TUMOR CELLS

Another type of tumor antigens are the proteins with normal level on non-malignant tissues and overexpressed status on the malignant tissues. There are several overexpressed tumor antigens including p53, human telomerase reverse transcriptase (hTERT), Epidermal growth factor receptor (EGFR), carbonic anhydrase IX. They usually play a central role in in developing malignancy, therefore; the cancerous cells do not simply reduce them to escape from immune recognition. The efficacy and safety resulted by targeting this class of TAA depends on expression differences between malignant and normal tissue expression.

A novel method has been presented recently that uses mini-genes to identify immunogenic tumor-associated mutations. In this approach, whole exome sequencing was done on normal and tumor cell

DNA in order to identify nonsynonymous mutations. The flanking 12 amino acids containing mutated amino acids are encoded by a 75-nucleotide mini-gene. Then combination of 8–15 different mini-genes are gathered and concatenated into longer tandem mini-gene constructs. Cloning of the constructs are carried out through expression vector. Transcription as RNA, and electroporation into the autologous dendritic cells (DCs) of patients are the next steps to show the 8–15 candidate epitopes on DCs in the context of all of the MHC loci of patients. Then TILs are added and evaluated for reactivity. Lu et al. showed that implementing this technique would be able to rediscover T cells in TIL of melanoma and lead to recognition of tumor-specific neoantigens and subsequently make feasible the discovery of new mutated antigens, which are not found through standard expression cloning.[47]

EPITOPE SPREADING

Targeting specific tumor antigens lead to limited autoimmunity and developing a very specific effector and memory cell response. Such a method may biologically have opposite effects. This might lead to increasing the frequency of tumor cells which lack the targeted tumor antigen, therefore; they are resistant to vaccine therapy. However, the resistant tumors might have an advantage of expressing new antigens and then expand the immune response against the tumor, i.e. developing immune response against the antigens not included in the vaccine. The phenomenon is named epitope spreading. The term *epitope spreading*, also known as determinant spreading, is defined as the spread of the immune response from one antigen to another antigen expressed in the same tissue. This spontaneous condition has been studied in autoimmune diseases and has also been noted in cancer immunotherapy studies.[48–51]

The phenomenon of releasing tumor antigens as a result of T cell attack to tumor cells and lysing them to promote subsequent waves of T cells directed against different antigens can be a substantial occurrence for tumor rejection.[52] To provide episodes of immunity targeting unidentified, patient-specific mutated antigens, which can lead to tumor rejection, vaccines that target shared antigens can be beneficial. Several clinical trials have shown that lead to immune responses against the vaccine-targeted antigen is not correlated with the clinical outcome. This may be related to antigen spreading and developing immune responses against other tumor antigens which have not been measured.[48] The significant role of TAA-containing vaccines may be to trigger epitope spread to tumor-specific antigens that activate T cells with higher avidity, which mediates tumor rejection more effectively.[53]

TUMOR-ASSOCIATED ANTIGENS

TAAs are not specifically expressed on tumors and can be present on normal cells as well. They also fall into the category of shared antigens because they are shared between various types of tumors.[54,55]

TAAs show significant variability in their immunogenicity, and to escape immune recognition they may undergo immune editing and may have different expression patterns between tumor types and especially between individuals.[56]

TAAs can be categorized based on their expression pattern in healthy tissues. First group is the overexpressed antigens.[57,58] Expression levels of these antigens are higher in tumor cells compared with tumor cells. The second group is the differentiation antigens which are specific for a cell lineage such as the melanocyte differentiation antigens that are expressed by the most melanoma tumors.[59,60]

However, the currently accepted classification of cancer antigens is based on their shared and unique attribute and further, regarding the HLA molecules, they can be categorized as class I and class II HLA-restricted antigens.[61] Shared TAAs, as it is obvious, are not specific for a kind of tumor and are detectable within normal tissues of many individuals. Altogether, shared TAAs include two groups, differentiation antigens (such as Gp100, Melan-A/Mart-1, tyrosinase, PSA, CEA, and Mammaglobin-A), and overexpressed antigens (such as p53, surivivin, HER2/neu).[62] Other examples of shared TAAs are hTERT,[63] preferentially expressed antigen of melanoma (PRAME),[64] proteinase-3,[65,66] survivin,[67,68] Wilms tumor gene-encoded transcription factor-1 (WT1),[69,70] and mucin-1.[71] Potential targets for cancer immunotherapy, shared TAAs have been become the center of attention. In a study by Arai et al., cytotoxicity was induced by CD8+ CTLs specific for hTERT peptides against leukemia cells in an HLA-A24–restricted manner[72] PRAME is a TAA with at least four various HLA-A*0201-restricted epitopes which were recognized by CTLs.[58,73] PRAME antigens have been reported positive in 47%–70% of patients with acute myeloid leukemia (AML).[74,75] Proteinase-3 is also overexpressed in AML and chronic myeloid leukemia (CML) and WT1 is noticeably present in many types of leukemia.[54] Based on the reports of the National Cancer Institute about the prioritization of cancer antigens, WT1 has been considered among the top 20 antigens, which are best candidates for cancer immunotherapy.[76] Furthermore, WT1 an appropriate target for AML immunotherapy and have been tested in several vaccine trials for AML treatment.

TUMOR-SPECIFIC ANTIGENS

TSAs are antigens which are only expressed on the malignant cells and can stimulate immune responses which specifically attack the tumor cells. Unique TSAs are only present on a certain tumor and might not expressed by any other tumors of that type. Because they are developed through random somatic point mutations, they are expressed uniquely by individual tumors and even may be not by other parts of the tumor.[77] These TSAs display attributes of each single tumor and are diverse between malignancies developed in the same animal. Even they might be diverse in other parts of the same tumoral nodule.[78–80] If the mutated antigen is a oncoantigen, which plays a central role in developing malignant phenotype of the tumor, it would not be a case of immune escape. Unique antigens are usually candidates of designing more effective immunotherapies.[62] This aim has been achieved by utilizing autologous tumor cells,[81] tumor lysates,[82] autologous tumor-derived HSPs,[83] DCs pulsed with autologous tumor cells or tumor-derived RNA.[84] Unique antigens revealed improved clinical outcomes in renal cancer patients receiving GM-CSF secreting tumor cells[81] and patients with melanoma. Lurquin et al. reported the effects of antigen spreading on the melanoma patients received MAGE-A3 vaccine, who generated clonal T cell expansion toward antigens, which were not targeted by the vaccine but were present on the patient's tumor cells, specially unique antigens.[85]

Shared TSAs are tumor specific but shared between various malignant cells. For example mutation in tumor suppressor genes or oncogenes results in expression of TSAs shared between many types of tumors. Another example of these TSAs are cancer/testis (CT) antigens, which are proteins expressed by a few healthy tissues, i.e. the germline cells and placenta. However, because of dysregulation of the demethylation in tumors cells these antigens can be expressed by various types of malignant cells as well, and deserve as shared TSAs.[54,86,87] Heavy and light chains of immunoglobulins (Ig) at the amino termini have highly specific variable regions. The variable regions of both chains together make the unique antigen-recognition area of the Ig protein. The variable regions have determinants which can act as an antigen or idiotypes (Ids), therefore, in B cell malignancies expressing only one type of Igs, Ids can be recognized as TSAs. The desired tumor antigen is one present selectively on the malignant cells, expressed by all tumors of that type, vital for survival of malignant cells, and capable to provoke both humoral and cell-mediated immunity.[88] It is demonstrated that idiotype protein triggers strong humoral immune response because anti-idiotype antibodies could easily be detected in animals or patients who had undergone vaccination.[89] This was feasible after discovery of the earliest TSAs in the 1970s.[90,91] Idiotype vaccines are polyclonal, and lead to inducing immune responses against multiple epitopes of a tumor antigen.

In general, this type of tumor antigen may be a result of expression of viral antigens or abnormal proteins that stem from somatic mutations in malignant cells, which are neoantigens Driver mutations accounting for the progression of tumors and passenger mutations which are byproduct of genetic instability might lead to expression of TSAs.[92] Immunonologic tolerance are less probable toward the antigens originated from mutations.[93,94]

NEOANTIGENS

Mutated genes encode targets that remove issues stemming from expression in normal tissue and that can potentially be presented in almost all cancers except some hematological malignancies that carry fewer than 10 nonsynonymous somatic mutations. Recent studies show that recognition of mutated tumor antigens, which are also called neoantigens, that correlated with clinical responses in a variety of therapeutic settings have provided a motivation to design novel therapeutic approaches based upon analysis of the mutational landscape of tumors.[95,96] In 2005, two major studies on humans raised enthusiasm concerning tumor neoantigens as therapeutic targets for cancer immunotherapy. The first one, carried out by T. Wölfel et al. demonstrated that expression-cloning methods in a melanoma patient resulted in developin antitumor T cell response toward neoantigens, made by somatic point mutations in five distinct genes. This displayed that T cell responses against these TSAs overcame responses toward TAAs.[97] Second, Rosenberg and Robbins demonstrated that TILs, expanded *ex vivo* and adoptively administered to a patient with melanoma who showed a complete tumor regression, was consisted of T cells specific for two mutant tumor antigens.[98]

In 2012, for the first time it was recorded that genomics and bioinformatics methods have the potential to identify tumor-specific mutant antigens that act as TSAs. Matsushita et al. and Sahin et al. by combining next-generation sequencing, computer-based prediction of epitope, and immunological evaluation identified and validated distinct TSAs in highly immunogenic methylcholanthrene-induced sarcoma cells and in melanoma tumor cells of mice, respectively.[99,100]

The following year, Robbins and Rosenberg showed that by using whole exome sequencing on human tumors, neoantigens, which were recognized by adoptively transferred tumor-reactive T cells, were identified.[101] Together, the investigations on mouse and human indicated the possibility of using genomics and bioinformatics tools to identify tumor neoantigens.[92]

The rate of mutation varies significantly from cancer to cancer and even within different cancer types.[102–104] The variations of nonsynonymous mutations range from about 10 mutations per genome in AML to hundreds of mutations per genome in cancers such as lung cancer and melanoma.[105–107] Theoretically, filtering this limited number of coding mutations through peptide-MHC affinity prediction algorithms would yield a manageable final number of potential neoantigens per tumor.[108]

Various clinical implications stem from the identification of specific neoantigens. Identifying neoantigens provides a target to design tumor specific vaccines. Theoretically, neoantigen-derived vaccines would be less susceptible to immune tolerance than currently used tumor vaccines, and it could be easy to measure the immune responses to vaccines. Moreover, understanding of specific neoantigens could assist with several challenges facing the next generation of immunomodulatory agents in development, such as antibodies targeting CD137 and PD1/PD-L1. Although in clinical trials these agents have generated some considerable outcomes, responses have been limited to a subset of patients without distinct biomarkers identified to help in predicting response. Understanding the presence of tumor-specific T cells and the immunophenotype of these tumor-specific T cells may help identify those patients most likely to benefit from immunomodulatory therapy. For instance, it is known that expression of PDL1 by tumor cells does not accurately predict response to immune checkpoint blockade,[109] showing that perhaps the phenotype of the tumor-specific lymphocytes may be a better predictor than the tumor itself. One possibility is that expression of PD-1 on neoantigen-specific T cells may better predict response to anti-PD-1/PD-L1 antibody therapy. In addition, after immunomodulatory therapy, monitoring the neoantigen-specific tumor response may provide a better evaluation of response than conventional methods of imaging, which are unable to differentiate progressive disease from early immune responses (pseudoprogression). Finally, identification of neoantigen-specific lymphocytes could ameliorate the production protocols used for adoptive immunotherapy from TILs. As an alternative method to reversing T cell depletion with inhibitors of immune checkpoint, it is possible to easily clone TCRs from neoantigen-specific cells and transduce them into lymphocytes. There is evidence that transduction of tumor-specific TCRs into naive cells may produce a more effective population of T cells for adoptive immunotherapy.[110]

GLYCOLIPIDS AND GLYCOPROTEINS AS ANTIGENS

Malignant cells contain carbohydrates on their surface, which are undergone alterations by glycosyltransferases glycosidase and are not similar to the normal molecules on healthy cells. Carbohydrate structure in tumor cells changes mostly quantitatively and not qualitatively. This term means that to compare with normal cells, a smaller quantity of the carbohydrate's structure is observed in tumor cells. Despite that, the structure of cell surface carbohydrates on tumor cells may be so diverse that as tumor-associated antigens, they can be recognized by immunological approaches. Carbohydrate antigens may play an important role in the cell recognition process during development and differentiation. Two general approaches have been utilized to identify the tumor-associated carbohydrate antigens (TACAs). The first approach comprises a systematic chemical characterization of carbohydrate composition and structure, and if such a structure is detected it leads to antibody production to an abnormal component. The second approach is evolved from tumor-specific monoclonal antibodies that are directed to carbohydrate antigens. By using this technique, a number of glycolipid antigens have been identified in a variety of human cancers.[111—113]

Two groups of carbohydrates, which are different in chemical structure, demonstrate extreme changes in transformed cells or tumor cells: (1) carbohydrates, which join to ceramides (glycosphingolipids or glycolipids); and (2) carbohydrates that are adhered to the proteins on the surface of cells (glycoproteins).[111] Until now, different TACAs have been identified. These include: (1) the mucin-related TACAs such as Thomsen-Friendreich (TF), Thomsen-nouveau (Tn), and sialyl-Tn

(STn), which are attached to membrane-bound mucin proteins through α-O-Ser/Thr linkages and associated with epithelial cancers; (2) the blood group Lewis-related TACAs such as Lewis^X (SSEA-1), sialyl Lewis^X, Lewis^Y, and sialyl Lewis^a, which are glycolipids showing overexpression in some types of cancers such as breast, ovary, prostate, liver, and colon; (3) the Globo class such as Globo-H, Gb3, Gb4, Gb5 (SSEA-3), which showed upregulation in the malignant tissues similar to the Lewis-related antigens and (4) the gangliosides such as GD2, GD3, GM2, GM3, GT1b, RM2, A2B5 (c-series gangliosides), fucosyl-GM1, and polysialic acid (PSA), upregulated in renal, colon, lung, and prostate cancers.[114–121] It has been realized that the stage-specific embryonic antigen 4 (SSEA-4) became a candidate target for treatment of astrocytoma.[121]

TACAs are thought to play a role in the progression and metastasis of malignant cells. The upregulation of TACAs on surface of malignant cells may be an indicator of poor prognosis.[122–125] Furthermore, various malignant cells may share the same type of TACAs and may be present as a type of TACA in the surface of a wide range of various tumor cells. In this regard, designing a broadly effective TACA-based vaccine is possible, which may be appropriate for a range of cancer types and represent promising candidates for cancer immunotherapy.[126]

MONO-EPITOPE VERSUS POLY-EPITOPE ANTIGEN

For the most specific method to trigger T cells against cancer cells, epitope-based cancer vaccines were used and investigated by employing peptide-based vaccination for the first time. There are findings that support this approach: (1) a T cell recognizes only a single epitope (peptide) bound to MHC proteins within a given antigen; (2) MHC class I and class II molecules attach, respectively, to short peptides of 8–10 and 13–20 amino acids, (3) MHC class I- and class II-presented peptides activate CD4+ and CD8+ T cells, respectively, and (4) TAAs are present on cancer cells.[127]

Although, mono-epitope cancer vaccines have produced somewhat poor clinical outcomes in patients with cancer, the development of novel altered peptides and the understanding of the molecular immune regulatory mechanisms of cancer cells have provided new promising use of mono-epitope peptide-based vaccination in cancer patients.[127]

The principle for the first application of TAA poly-epitope cancer vaccines was based on the findings that: (1) the administration of multiple epitopes can prevail the potential down regulation of a administered TAA-single epitope in tumor cells in comparison to a mono-epitope-based cancer vaccine, (2) poly-epitope cancer vaccines, having capability to target more than one TAA, can prevent the heterogeneous expression of TAAs by different malignant cells within a tumor, and (3) by using poly-epitope vaccines, the criteria of eligibility for enrollment in a clinical trial would not be restricted to the patients with a certain HLA.[128,129] Subsequently, poly-epitope cancer vaccines might overcome the variation of immunodominant peptides presentation by malignant cells, and also by engineering the peptides, they can induce *in vivo* activation of different high avidity TCR repertoires specific for self TAA peptides. The feasibility and tolerability of vaccination with a combination of 11 synthetic peptides stemmed from a number of TAAs, including prostate-specific antigen (PSA) and prostate-specific membrane antigens (PSMA), was evaluated in a phase I/II trial in 19 HLA-A2 positive patients with hormone-sensitive prostate carcinoma. The result of this trial showed that in 4 of 19 cases, PSA progress was stable or slowed down by the poly-peptide vaccine.[130] In another randomized clinical trial, of 38 patients with high-risk, resected stage III or IV melanoma, 33 showed an immune response after vaccination with two tumor antigen epitope peptides originated from gp100 and tyrosinase with incomplete Freund's adjuvant administered alone or in combination with IL-12.[131]

REFERENCES

1. Zerbini A, et al. Radiofrequency thermal ablation of hepatocellular carcinoma liver nodules can activate and enhance tumor-specific T-cell responses. *Cancer Res* 2006;**66**(2):1139–46.
2. Zitvogel L, Kepp O, Kroemer G. Immune parameters affecting the efficacy of chemotherapeutic regimens. *Nat Rev Clin Oncol* 2011;**8**(3):151.
3. Lurquin C, et al. Structure of the gene of tum– transplantation antigen P91A: the mutated exon encodes a peptide recognized with Ld by cytolytic T cells. *Cell* 1989;**58**(2):293–303.
4. Rammensee H-G, Falk K, Rötzschke O. Peptides naturally presented by MHC class I molecules. *Annu Rev Immunol* 1993;**11**(1):213–44.
5. Vigneron N, Van den Eynde BJ. Insights into the processing of MHC class I ligands gained from the study of human tumor epitopes. *Cell Mol Life Sci* 2011;**68**(9):1503–20.
6. Gross L. Intradermal immunization of C3H mice against a sarcoma that originated in an animal of the same line. *Cancer Res* 1943;**3**(5):326–33.
7. Foley E. Antigenic properties of methylcholanthrene-induced tumors in mice of the strain of origin. *Cancer Res* 1953;**13**(12):835–7.
8. Prehn RT, Main JM. Immunity to methylcholanthrene-induced sarcomas. *J Natl Cancer Inst* 1957;**18**(6): 769–78.
9. Old LJ. Cancer immunology: the search for specificity. *Natl Cancer Inst Monogr* 1982;**60**:193–209.
10. Bjorkman PJ, et al. Structure of the human class I histocompatibility antigen, HLA-A2. *Nature* 1987; **329**(6139):506.
11. Babbitt BP, et al. Binding of immunogenic peptides to Ia histocompatibility molecules. *Nature* 1985; **317**(6035):359.
12. Gillis S, Smith KA. Long term culture of tumour-specific cytotoxic T cells. *Nature* 1977;**268**(5616):154.
13. Cerottini J-C, et al. Generation of cytotoxic T lymphocytes in vitro: I. Response of normal and immune mouse spleen cells in mixed leukocyte cultures. *J Exp Med* 1974;**140**(3):703–17.
14. De Plaen E, et al. Immunogenic (tum-) variants of mouse tumor P815: cloning of the gene of tum-antigen P91A and identification of the tum-mutation. *Proc Natl Acad Sci U S A* 1988;**85**(7):2274–8.
15. Fridman WH, et al. The immune contexture in human tumours: impact on clinical outcome. *Nat Rev Cancer* 2012;**12**(4):nrc3245.
16. Erdag G, et al. Immunotype and immunohistologic characteristics of tumor-infiltrating immune cells are associated with clinical outcome in metastatic melanoma. *Cancer Res* 2012;**72**(5):1070–80.
17. Galon J, et al. Type, density, and location of immune cells within human colorectal tumors predict clinical outcome. *Science* 2006;**313**(5795):1960–4.
18. Zhang L, et al. Intratumoral T cells, recurrence, and survival in epithelial ovarian cancer. *N Engl J Med* 2003;**348**(3):203–13.
19. Galon J, et al. Towards the introduction of the 'Immunoscore'in the classification of malignant tumours. *J Pathol* 2014;**232**(2):199–209.
20. Pages F, et al. In situ cytotoxic and memory T cells predict outcome in patients with early-stage colorectal cancer. *J Clin Oncol* 2009;**27**(35):5944–51.
21. Tumeh PC, et al. PD-1 blockade induces responses by inhibiting adaptive immune resistance. *Nature* 2014; **515**(7528):568.

22. Johnson LA, et al. Gene therapy with human and mouse T-cell receptors mediates cancer regression and targets normal tissues expressing cognate antigen. *Blood* 2009;**114**(3):535–46.
23. Kawakami Y, et al. Identification of new melanoma epitopes on melanosomal proteins recognized by tumor infiltrating T lymphocytes restricted by HLA-A1,-A2, and-A3 alleles. *J Immunol* 1998;**161**(12):6985–92.
24. Germeau C, et al. High frequency of antitumor T cells in the blood of melanoma patients before and after vaccination with tumor antigens. *J Exp Med* 2005;**201**(2):241–8.
25. van der Bruggen P, et al. A gene encoding an antigen recognized by cytolytic T lymphocytes on a human melanoma. *Science* 1991;**254**(5038):1643–7.
26. Vigneron N, et al. Identifying source proteins for MHC class I-presented peptides. In: *Antigen Processing*. Springer; 2013. p. 187–207.
27. Brichard V, et al. The tyrosinase gene codes for an antigen recognized by autologous cytolytic T lymphocytes on HLA-A2 melanomas. *J Exp Med* 1993;**178**(2):489–95.
28. Ma W, et al. Two new tumor-specific antigenic peptides encoded by gene MAGE-C2 and presented to cytolytic T lymphocytes by HLA-A2. *Int J Cancer* 2004;**109**(5):698–702.
29. Ma W, et al. A MAGE-C2 antigenic peptide processed by the immunoproteasome is recognized by cytolytic T cells isolated from a melanoma patient after successful immunotherapy. *Int J Cancer* 2011;**129**(10): 2427–34.
30. Vigneron N, et al. Database of T cell-defined human tumor antigens: the 2013 update. *Cancer Immun Arch* 2013;**13**(3):15.
31. Vigneron N, et al. A peptide derived from melanocytic protein gp100 and presented by HLA-B35 is recognized by autologous cytolytic T lymphocytes on melanoma cells. *Tissue Antigens* 2005;**65**(2):156–62.
32. Tomita Y, et al. A novel tumor-associated antigen, cell division cycle 45-like can induce cytotoxic T-lymphocytes reactive to tumor cells. *Cancer Sci* 2011;**102**(4):697–705.
33. Hunt DF, et al. Characterization of peptides bound to the class I MHC molecule HLA-A2. 1 by mass spectrometry. *Science* 1992;**255**(5049):1261–3.
34. Henderson RA, et al. HLA-A2. 1-associated peptides from a mutant cell line: a second pathway of antigen presentation. *Science* 1992;**255**(5049):1264–6.
35. Cox AL, et al. Identification of a peptide recognized by five melanoma-specific human cytotoxic T cell lines. *Science* 1994;**264**(5159):716–9.
36. Schirle M, et al. Identification of tumor-associated MHC class I ligands by a novel T cell-independent approach. *Eur J Immunol* 2000;**30**(8):2216–25.
37. Vigneron N. Human tumor antigens and cancer immunotherapy. *BioMed Res Int* 2015;**2015**.
38. Skipper JC, et al. Mass-spectrometric evaluation of HLA-A* 0201-associated peptides identifies dominant naturally processed forms of CTL epitopes from MART-1 and gp100. *Int J Cancer* 1999;**82**(5):669–77.
39. Krüger T, et al. Lessons to be learned from primary renal cell carcinomas: novel tumor antigens and HLA ligands for immunotherapy. *Cancer Immunol Immunother* 2005;**54**(9):826–36.
40. Sahin U, et al. Human neoplasms elicit multiple specific immune responses in the autologous host. *Proc Natl Acad Sci U S A* 1995;**92**(25):11810–3.
41. Chen Y-T, et al. A testicular antigen aberrantly expressed in human cancers detected by autologous antibody screening. *Proc Natl Acad Sci U S A* 1997;**94**(5):1914–8.
42. Jungbluth AA, et al. Monophasic and biphasic synovial sarcomas abundantly express cancer/testis antigen NY-ESO-1 but not MAGE-A1 or CT7. *Int J Cancer* 2001;**94**(2):252–6.
43. Barrow C, et al. Tumor antigen expression in melanoma varies according to antigen and stage. *Clin Cancer Res* 2006;**12**(3):764–71.
44. Gure AO, et al. Cancer-testis genes are coordinately expressed and are markers of poor outcome in non–small cell lung cancer. *Clin Cancer Res* 2005;**11**(22):8055–62.

45. Robbins PF, et al. A pilot trial using lymphocytes genetically engineered with an NY-ESO-1–reactive T-cell receptor: long-term follow-up and correlates with response. *Clin Cancer Res* 2015;**21**(5):1019–27.
46. Novellino L, Castelli C, Parmiani G. A listing of human tumor antigens recognized by T cells: March 2004 update. *Cancer Immunol Immunother* 2005;**54**(3):187–207.
47. Lu Y-C, et al. Efficient identification of mutated cancer antigens recognized by T cells associated with durable tumor regressions. *Clin Cancer Res* 2014;**20**(13):3401–10.
48. Butterfield LH, et al. Determinant spreading associated with clinical response in dendritic cell-based immunotherapy for malignant melanoma. *Clin Cancer Res* 2003;**9**(3):998–1008.
49. Ranieri E, et al. Dendritic cell/peptide cancer vaccines: clinical responsiveness and epitope spreading. *Immunol Invest* 2000;**29**(2):121–5.
50. Lehmann PV, et al. Spreading of T-cell autoimmunity to cryptic determinants of an autoantigen. *Nature* 1992;**358**(6382):155.
51. Ribas A, et al. Role of dendritic cell phenotype, determinant spreading, and negative costimulatory blockade in dendritic cell-based melanoma immunotherapy. *J Immunother* 2004;**27**(5):354–67.
52. Disis ML. Immunologic biomarkers as correlates of clinical response to cancer immunotherapy. *Cancer Immunol Immunother* 2011;**60**(3):433–42.
53. Krauze MT, et al. Prognostic significance of autoimmunity during treatment of melanoma with interferon. In: *Seminars in immunopathology.* Springer; 2011.
54. Zilberberg J, Feinman R, Korngold R. Strategies for the Identification of T Cell–recognized tumor antigens in hematological malignancies for improved graft-versus-tumor responses after allogeneic blood and marrow transplantation. *Biol Blood Marrow Transplant* 2015;**21**(6):1000–7.
55. Ilyas S, Yang JC. Landscape of Tumor Antigens in T-Cell Immunotherapy. *Immunol (Baltimore, Md. : 1950)* 2015;**195**(11):5117–22. https://doi.org/10.4049/jimmunol.1501657.
56. Escors D. Tumour immunogenicity, antigen presentation, and immunological barriers in cancer immunotherapy. *New J Sci* 2014;**2014**.
57. Fisk B, et al. Identification of an immunodominant peptide of HER-2/neu protooncogene recognized by ovarian tumor-specific cytotoxic T lymphocyte lines. *J Exp Med* 1995;**181**(6):2109–17.
58. Kessler JH, et al. Efficient identification of Novel HLA-A* 0201–presented cytotoxic T lymphocyte epitopes in the widely expressed tumor antigen PRAME by proteasome-mediated digestion analysis. *J Exp Med* 2001;**193**(1):73–88.
59. Coulie PG, et al. A new gene coding for a differentiation antigen recognized by autologous cytolytic T lymphocytes on HLA-A2 melanomas. *J Exp Med* 1994;**180**(1):35–42.
60. Kawakami Y, et al. Identification of a human melanoma antigen recognized by tumor-infiltrating lymphocytes associated with in vivo tumor rejection. *Proc Natl Acad Sci U S A* 1994;**91**(14):6458–62.
61. Renkvist N, et al. A listing of human tumor antigens recognized by T cells. *Cancer Immunol Immunother* 2001;**50**(1):3–15.
62. Buonaguro L, et al. Translating tumor antigens into cancer vaccines. *Clin Vaccine Immunol* 2011;**18**(1): 23–34.
63. Hernández J, et al. Identification of a human telomerase reverse transcriptase peptide of low affinity for HLA A2. 1 that induces cytotoxic T lymphocytes and mediates lysis of tumor cells. *Proc Natl Acad Sci U S A* 2002;**99**(19):12275–80.
64. Matsushita M, et al. Preferentially expressed antigen of melanoma (PRAME) in the development of diagnostic and therapeutic methods for hematological malignancies. *Leuk Lymphoma* 2003;**44**(3):439–44.
65. Knights A, et al. A novel MHC-associated proteinase 3 peptide isolated from primary chronic myeloid leukaemia cells further supports the significance of this antigen for the immunotherapy of myeloid leukaemias. *Leukemia* 2006;**20**(6):1067.

66. Fujiwara H, et al. Identification and in vitro expansion of CD4+ and CD8+ T cells specific for human neutrophil elastase. *Blood* 2004;**103**(8):3076–83.
67. Friedrichs B, et al. Survivin-derived peptide epitopes and their role for induction of antitumor immunity in hematological malignancies. *Leuk Lymphoma* 2006;**47**(6):978–85.
68. Andersen M. Survivin–a universal tumor antigen. *Histol Histopathol* 2002;**17**(2):669–75.
69. Oka Y, et al. Human cytotoxic T-lymphocyte responses specific for peptides of the wild-type Wilms' tumor gene (WT1) product. *Immunogenetics* 2000;**51**(2):99–107.
70. Oka Y, et al. Cancer immunotherapy targeting Wilms' tumor gene WT1 product. *J Immunol* 2000;**164**(4): 1873–80.
71. Apostolopoulos V, McKenzie IF. Cellular mucins: targets for immunotherapy. *Crit Rev Immunol* 1994; **14**(3–4).
72. Arai J, et al. Identification of human telomerase reverse transcriptase–derived peptides that induce HLA-A24–restricted antileukemia cytotoxic T lymphocytes. *Blood* 2001;**97**(9):2903–7.
73. Rezvani K, et al. Ex vivo characterization of polyclonal memory CD8+ T-cell responses to PRAME-specific peptides in patients with acute lymphoblastic leukemia and acute and chronic myeloid leukemia. *Blood* 2009;**113**(10):2245–55.
74. Greiner J, et al. Expression of tumor-associated antigens in acute myeloid leukemia: implications for specific immunotherapeutic approaches. *Blood* 2006;**108**(13):4109–17.
75. Greiner J, et al. Simultaneous expression of different immunogenic antigens in acute myeloid leukemia. *Exp Hematol* 2000;**28**(12):1413–22.
76. Cheever MA, et al. The prioritization of cancer antigens: a national cancer institute pilot project for the acceleration of translational research. *Clin Cancer Res* 2009;**15**(17):5323–37.
77. Parmiani G, et al. Unique human tumor antigens: immunobiology and use in clinical trials. *J Immunol* 2007; **178**(4):1975–9.
78. Gilboa E. The makings of a tumor rejection antigen. *Immunity* 1999;**11**(3):263–70.
79. Prehn RT. Analysis of antigenic heterogeneity within individual 3-methylcholanthrene-induced mouse sarcomas. *J Natl Cancer Inst* 1970;**45**(5):1039–45.
80. Wortzel RD, Philipps C, Schreiber H. Multiple tumour-specific antigens expressed on a single tumour cell. *Nature* 1983;**304**(5922):165.
81. Mautner J, et al. Diverse CD8 T-cell responses to renal cell carcinoma antigens in patients treated with an autologous granulocyte-macrophage colony-stimulating factor gene-transduced renal tumor cell vaccine. *Cancer Res* 2005;**65**(3):1079–88.
82. Jocham D, et al. Adjuvant autologous renal tumour cell vaccine and risk of tumour progression in patients with renal-cell carcinoma after radical nephrectomy: phase III, randomised controlled trial. *Lancet* 2004; **363**(9409):594–9.
83. Castelli C, et al. Heat shock proteins: biological functions and clinical application as personalized vaccines for human cancer. Cancer immunology. *Immunotherapy* 2004;**53**(3):227–33.
84. Su Z, et al. Immunological and clinical responses in metastatic renal cancer patients vaccinated with tumor RNA-transfected dendritic cells. *Cancer Res* 2003;**63**(9):2127–33.
85. Lurquin C, et al. Contrasting frequencies of antitumor and anti-vaccine T cells in metastases of a melanoma patient vaccinated with a MAGE tumor antigen. *J Exp Med* 2005;**201**(2):249–57.
86. Chomez P, et al. An overview of the MAGE gene family with the identification of all human members of the family. *Cancer Res* 2001;**61**(14):5544–51.
87. Sensi M, Anichini A. Unique Tumor Antigens. Evidence for immune control of genome integrity and immunogenic targets for T cell mediated patient-specific immunotherapy. *Clin Cancer Res* 2006;**12**(17): 5023–32. https://doi.org/10.1158/1078-0432.CCR-05-2682.

88. Baio FE, Kwak LW, Weng J. Towards an off-the-shelf vaccine therapy targeting shared B-cell tumor idiotypes. *Am J Blood Res* 2014;**4**(2):46.
89. Dols A, et al. Allogeneic breast cancer cell vaccines. *Clin Breast Cancer* 2003;**3**(Suppl. 4):S173−80.
90. Sirisinha S, Eisen HN. Autoimmune-like antibodies to the ligand-binding sites of myeloma proteins. *Proc Natl Acad Sci U S A* 1971;**68**(12):3130−5.
91. Lynch RG, et al. Myeloma proteins as tumor-specific transplantation antigens. *Proc Natl Acad Sci U S A* 1972;**69**(6):1540−4.
92. Gubin MM, et al. Tumor neoantigens: building a framework for personalized cancer immunotherapy. *J Clin Invest* 2015;**125**(9):3413−21.
93. Heemskerk B, Kvistborg P, Schumacher TN. The cancer antigenome. *EMBO J* 2013;**32**(2):194−203.
94. Schumacher TN, Schreiber RD. Neoantigens in cancer immunotherapy. *Science* 2015;**348**(6230):69−74.
95. Watson IR, et al. Emerging patterns of somatic mutations in cancer. *Nat Rev Genet* 2013;**14**(10):703.
96. Lu YC, Robbins PF. Targeting neoantigens for cancer immunotherapy. *Int Immunol* 2016;**28**(7):365−70.
97. Lennerz V, et al. The response of autologous T cells to a human melanoma is dominated by mutated neoantigens. *Proc Natl Acad Sci U S A* 2005;**102**(44):16013−8.
98. Zhou J, et al. Persistence of multiple tumor-specific T-cell clones is associated with complete tumor regression in a melanoma patient receiving adoptive cell transfer therapy. *J Immunother* 2005;**28**(1):53.
99. Matsushita H, et al. Cancer exome analysis reveals a T-cell-dependent mechanism of cancer immunoediting. *Nature* 2012;**482**(7385):400.
100. Castle JC, et al. Exploiting the mutanome for tumor vaccination. *Cancer Res* 2012;**72**(5):1081−91.
101. Robbins PF, et al. Mining exomic sequencing data to identify mutated antigens recognized by adoptively transferred tumor-reactive T cells. *Nat Med* 2013;**19**(6):747.
102. Vogelstein B, et al. Cancer genome landscapes. *Science* 2013;**339**(6127):1546−58.
103. Kandoth C, et al. Mutational landscape and significance across 12 major cancer types. *Nature* 2013; **502**(7471):333.
104. Lawrence MS, et al. Mutational heterogeneity in cancer and the search for new cancer-associated genes. *Nature* 2013;**499**(7457):214.
105. Govindan R, et al. Genomic landscape of non-small cell lung cancer in smokers and never-smokers. *Cell* 2012;**150**(6):1121−34.
106. Network CGAR. Genomic and epigenomic landscapes of adult de novo acute myeloid leukemia. *N Engl J Med* 2013;**368**(22):2059−74.
107. Wei X, et al. Exome sequencing identifies GRIN2A as frequently mutated in melanoma. *Nat Genet* 2011; **43**(5):442.
108. Khodadoust MS, Alizadeh AA. Tumor antigen discovery through translation of the cancer genome. *Immunol Res* 2014;**58**(2−3):292−9.
109. Wolchok JD, et al. Nivolumab plus ipilimumab in advanced melanoma. *N Engl J Med* 2013;**369**(2):122−33.
110. Hinrichs CS, et al. Adoptively transferred effector cells derived from naive rather than central memory CD8+ T cells mediate superior antitumor immunity. *Proc Natl Acad Sci U S A* 2009;**106**(41):17469−74.
111. Hakomori S-i. Tumor-associated carbohydrate antigens. *Annu Rev Immunol* 1984;**2**(1):103−26.
112. Livingston PO. Approaches to augmenting the immunogenicity of melanoma gangliosides: from whole melanoma cells to ganglioside-KLH conjugate vaccines. *Immunol Rev* 1995;**145**(1):147−66.
113. Monzavi-Karbassi B, Pashov A, Kieber-Emmons T. Tumor-associated glycans and immune surveillance. *Vaccines* 2013;**1**(2):174−203.
114. Park C, Bergsagel D, McCulloch E. Mouse myeloma tumor stem cells: a primary cell culture assay. *J Natl Cancer Inst* 1971;**46**(2):411−22.
115. Komminoth P, et al. Polysialic acid of the neural cell adhesion molecule distinguishes small cell lung carcinoma from carcinoids. *Am J Pathol* 1991;**139**(2):297.

116. Kobata A, Amano J. Altered glycosylation of proteins produced by malignant cells, and application for the diagnosis and immunotherapy of tumours. *Immunol Cell Biol* 2005;**83**(4):429–39.
117. Balic M, et al. Most early disseminated cancer cells detected in bone marrow of breast cancer patients have a putative breast cancer stem cell phenotype. *Clin Cancer Res* 2006;**12**(19):5615–21.
118. Chang W-W, et al. Expression of Globo H and SSEA3 in breast cancer stem cells and the involvement of fucosyl transferases 1 and 2 in Globo H synthesis. *Proc Natl Acad Sci U S A* 2008;**105**(33):11667–72.
119. Cazet A, et al. Tumour-associated carbohydrate antigens in breast cancer. *Breast Cancer Res* 2010;**12**(3): 204.
120. Heimburg-Molinaro J, et al. Cancer vaccines and carbohydrate epitopes. *Vaccine* 2011;**29**(48):8802–26.
121. Lou Y-W, et al. Stage-specific embryonic antigen-4 as a potential therapeutic target in glioblastoma multiforme and other cancers. *Proc Natl Acad Sci U S A* 2014;**111**(7):2482–7.
122. Liu C-C, Ye X-S. Carbohydrate-based cancer vaccines: target cancer with sugar bullets. *Glycoconj J* 2012; **29**(5–6):259–71.
123. Fukuda M. Possible roles of tumor-associated carbohydrate antigens. *Cancer Res* 1996;**56**(10):2237–44.
124. Hakomori S-I, Zhang Y. Glycosphingolipid antigens and cancer therapy. *Chem Biol* 1997;**4**(2):97–104.
125. Werther JL, et al. Sialosyl-Tn, antigen as a marker of gastric cancer progression: an international study. *Int J Cancer* 1996;**69**(3):193–9.
126. Wei MM, Wang YS, Ye XS. Carbohydrate-based vaccines for oncotherapy. *Med Res Rev* 2018;**38**.
127. Bei R, Scardino A. TAA polyepitope DNA-based vaccines: a potential tool for cancer therapy. *BioMed Res Int* 2010;**2010**.
128. Pilla L, et al. Multipeptide vaccination in cancer patients. *Expet Opin Biol Ther* 2009;**9**(8):1043–55.
129. Parmiani G, et al. Cancer immunotherapy with peptide-based vaccines: what have we achieved? Where are we going? *J Natl Cancer Inst* 2002;**94**(11):805–18.
130. Feyerabend S, et al. Novel multi-peptide vaccination in Hla-A2+ hormone sensitive patients with biochemical relapse of prostate cancer. *Prostate* 2009;**69**(9):917–27.
131. Lee P, et al. Effects of interleukin-12 on the immune response to a multipeptide vaccine for resected metastatic melanoma. *J Clin Oncol* 2001;**19**(18):3836–47.

CHAPTER

STRATEGY OF ALLOGENEIC AND AUTOLOGOUS CANCER VACCINES

5

Saeed Farajzadeh Valilou[1], **Nima Rezaei**[2,3,4]

Cancer Immunology Project (CIP), Universal Scientific Education and Research Network (USERN), Tehran, Iran[1]; *Research Center for Immunodeficiencies, Children's Medical Center, Tehran University of Medical Sciences, Tehran, Iran*[2]; *Department of Immunology, School of Medicine, Tehran University of Medical Sciences, Tehran, Iran*[3]; *Network of Immunity in Infection, Malignancy and Autoimmunity (NIIMA), Universal Scientific Education and Research Network (USERN), Tehran, Iran*[4]

AUTOLOGOUS CANCER VACCINES

Autologous cancer vaccines have many theoretical advantages. Most importantly, this type of vaccine is likely to contain unique or rare tumor antigens that develop through mutational events. Autologous cancer vaccines are used for optimum antigen presentation to T cells and then, of course, appropriately HLA matched. To produce autologous cancer vaccines, sufficient tumor must be attainable. This might be practically challenging even in melanoma which is a skin tumor and is more accessible. Production of autologous vaccines is not feasible in patients, who are free of clinically present disease and come after removal surgery of the lesion. Moreover, a consensus should be developed for agreement on the methods for processing, preserving, modification, and delivery of the autologous vaccines. Vaccination schedule is important to achieve favorable results. However, amount of tumor available for production of autologous vaccines accounts for preparation of only two or three doses of vaccine. The delay made between harvesting the tumor and preparation of vaccine is another concern, which can restrict the efficacy of this type of vaccine.[1]

IN VIVO STUDIES AND CLINICAL IMPLICATIONS

Autologous melanoma cancer vaccines have shown clinical failure in many phase III clinical trials. Phase III clinical trials have been used for processing autologous tumor in order to extract heat shock proteins (HSPs), which provided promising results. These proteins within the cells function as "chaperones" for peptide antigens and therefore simply and potentially present tumor antigens to the immune system.[2,3] Furthermore, it is possible to quantify the extracted product, which has important advantages for quality assurance in securing regulatory approval. The study collected 322 patients out of 451 screened patients who had randomly undergone vaccine therapy versus physician's choice of therapy, which could be single-agent chemotherapy, interleukin 2, or complete resection. Yet only the

Vaccines for Cancer Immunotherapy. https://doi.org/10.1016/B978-0-12-814039-0.00005-9

preliminary results of this experiment are available and statistically present without any significant benefit (or detriment) for the heat shock vaccine compared with physician's choice of therapy. However, an interesting primary experiment is that the group of patients with M1a disease with manifestation of soft tissue, nodal, or skin metastasis with a normal serum lactate dehydrogenase (LDH) level and treated with the vaccine had promising results and increased overall survival (OS) than those receiving physician's choice of therapy. The limitations of this approach remain apparent beside the expected detailed and mature results of this phase III trial; only a few patients with melanoma have accessible tumor from which the heat shock protein autologous vaccine can be produced, the randomized clinical trial is being conducted in patients with advanced metastatic disease, and the vaccine is being compared with a variety of therapies with different efficacy.[1] Irradiated autologous tumor cells were administered combined with dinitrophenyl and Bacillus Calmette-Guerin (BCG) as a vaccine for 126 patients with stage III melanoma.[4] It has previously been shown that the addition of BCG to the autologous vaccine augments response to autologous vaccination protocols.[5,6] M. Lotem et al. showed that using this therapeutic approach led to a median OS of 88 months a 5-year survival of 54%. Delayed type hypersensitivity (DTH) response was developed in response to the vaccine on days 5 and 8. Patients with a powerful DTH response showed improved 5-year OS compared with the patients without powerful DTH (75% versus 44%, respectively). Gene expression array revealed association of a 50-gene signature and prognosis. The cancer-testis antigens including MAGEA1, NY-ESO-2, SSX1, and SSX4, are among these genes, whose expression reported as a prognostic marker only in the setting of autologous vaccination. In this study, 35 patients received ipilimumab following the vaccine, and showed the 3-year survival of 46% compared with 19% in patients received ipilimumab alone ($p = 0.007$).[4] Autologous tumor combined with BCG for cancer immunotherapy is used not only in melanoma but also in several types of cancers including breast cancer patients. Convit et al. demonstrated that the 5-year follow-up of immunotherapy of mice and human breast cancer, showed a survival rate of 60% for the treatment consisting of autologous vaccine combined with BCG and formalin. This treatment approach could be a possible and safe therapeutic strategy although it requires further studies to determine the efficacy.[7]

In a study, an immunotherapeutic site (ITS) using the autologous thermostable hemoderivative-cancer vaccine combined with granulocyte macrophage-colony stimulant factor (GM-CSF) and etoposide was examined for patients with progressive ovarian cancer GM-CSF and etoposide were administered at the same site with vaccination to modulate the cellularity in the draining lymph node. The lymph node cell populations were immunophenotyped and exhibited that ITS obtained a locally protective immune profile T-regulatory-cells/activated-antigen presenting-cells and systemically augmented the antiprogressive effect of the examined vaccine.[8] In another study on advanced non-small cell lung carcinoma patients, autologous tumor-derived autophagosome vaccine made from tumor cells of pleural effusions was used to assess the safety and immune response. Four patients were enrolled in this pilot study. Two cycles of docetaxel (75 mg/m^2) were administered on days 1 and 29 in order to benefit from direct and immunomodulatory effects of chemotherapy. Two of three patients developed autologous tumor-specific immune response and all 4 patients showed specific antibody responses In this small trial, autologous tumor-derived autophagosome vaccine in combination with GM-CSF was reported safe and capable of inducing an immune response against cancerous cells.[9,10]

Autologous cancer vaccines were also tested in glioblastoma patients. Glioblastoma is the most common primary brain cancer and is correlated with a generally poor prognosis and approximately

16 months of a median OS.[11,12] In a phase II clinical trial, combination of surgical removal, concurrent radiation therapy (RT), chemotherapy using temozolomide, and an autologous HSP vaccine, HSPPC-96 vaccine or Prophage, was tested in glioblastoma. This vaccine was prepared from resected tumors and after RT was administered to the patients intradermally and in a weekly schedule and after four doses of vaccine, chemotherapy plus the fifth dose of vaccine were administered. Next doses of vaccine were injected in a monthly schedule. For 46 patients who received the vaccine, the median OS was 23.8 months. In addition, immunosuppression induced by tumor and mediated by PD-L1 expression on tumor and circulating immune cells, possibly affected the efficacy of vaccination. PD-L1 expression on peripheral myeloid cells was also assessed for the first time as a survival predictor. Median OS for patients who highly expressed PD-L1 on their myeloid cells was 18.0 months as compared with 44.7 months for patients who had low expression of PD-L1 (hazard ratio for death in patients with high expression 3.3; 95% CI, 1.4–8.6; $p = 0.007$). In this regard, there might be a survival benefit for glioblastoma patients when treated with autologous tumor-derived heat shock protein combined with standard therapy. Also, the efficacy of vaccine might noticeably decrease due to systemic immunosuppression induced by peripheral myeloid PD-L1 expression.[13]

Apoptotic tumor cells (ATCs) and dendritic cells (DCs) also can be utilized to develop autologous vaccine, which were tested for patients with stage III/IV head and neck squamous cell carcinoma (HNSCC). In this study, Whiteside et al. prepared autologous DCs from monocytes, and ATCs were used as a source of tumor antigen. The vaccine was injected intranodally, to which DTH and immunological response were developed and evaluated prevaccination and postvaccination. They showed that five of 10 patients with sufficient sterile tumor cells were leukapheresed to generate DCs. In three of four patients, who received the vaccine ATC-reactive T cells were found. Moreover, all four patients survived with no clinical evidence of disease for more than 5 years. In sum, the study showed safety and capability of inducing immune response of the vaccine. However, only the patients with positive DTH before vaccination, who had sufficient amount of tumor cells ($\geq 10^7$) were eligible for this treatment.[14]

ALLOGENEIC CANCER VACCINES

Allogeneic cancer vaccines are made up of intact or modified cancer cells from other patients selected for the using shared antigens available on a large percentage of similar tumor types. They have noticeable advantages over autologous cancer vaccines in terms of availability for patients in all stages of the disease and provide the capability to administer multiple vaccinations over a protracted period. In comparison with autologous cancer vaccines, they may also be recognized more simply by the immune system of the patient. Also, they do not require harvesting of tumor, are easy to administer, and are significantly more immunogenic than autologous cells because they can induce graft versus host-like immune responses. On the other hand, unique or rare antigens maybe absent in this type of vaccine, which could be of interest as antigenic targets in any given patient's tumor. Although allogeneic tumor vaccines are amenable to a degree of standardization and can be generated in enough large quantities to provide large-scale randomized trials in multiple institutions, there remain significant issues in the production and standardization of the final vaccine product that have had a significant adverse effect on the commercialization of vaccines.[1,15,16]

IN VITRO/IN VIVO STUDIES AND CLINICAL IMPLICATIONS

One allogeneic cancer vaccine is Canvaxin, which has conquer most of the hurdles such as standardization and production. Canvaxin is a polyvalent allogeneic vaccine composed of three irradiated melanoma cell lines. Dr. Donald Morton was generated this vaccine and then established a company named CancerVax, which produced Canavaxin in partnership with Serono.[17,18] Two multiinstitutional randomized phase III trials assessed Canavaxin for patients with resected stage III and stage IV melanoma. In the year 2005, the Data Safety Monitoring Board (DSMB) observing and controlling these two clinical studies decided to cease the trials due to an interim analysis which did not revealed survival advantages for the vaccine compared with placebo. This was a tragic happening for everyone involved in melanoma therapy. Investigators can learn a lot from the negative points observed during these trials. Probably the challenges in theses studies are representative of the the hurdles in the way of developing an effective therapeutic vaccine for melanoma.[1]

Addition of the gene encoding GM-CSF can make a more powerful autologous and allogeneic tumor cell vaccines. However, using autologous vaccines for large-scale production is not practical and allogeneic vaccines are the better cases. GM-CSF genetic modification was applied to the melanoma cell line, K1735-M2 (H-2k), and an allogeneic vaccine was generated. In a preclinical study on C57BL/6J mice, the vaccine showed a specific antitumor response leading to protection against development of the syngeneic B16–F10 (H-2b) melanoma tumor *in vitro* studies of T cell showed that allogeneic cell vaccination of animals resulted in cytotoxic T cells specific for the autologous tumor. In addition, *in vivo* depletion of T cell subset demonstrated that this antitumor outcome was mediated by both CD4+ and CD8+ T cells, indicating that the allogeneic vaccine may function through cross-priming while tumor antigens are processed and presented to T cells by antigen presenting cells of the host. Therefore, genetic modification of K1735-M2 cells with a retroviral vector encoding GM-CSF led to an antitumor activity of the melanoma cell lines as a therapeutic vaccine against the syngeneic B16–F10 tumor.[19]

In addition to the melanoma, allogeneic cancer vaccines are used in other type of cancers such as bladder, lung, prostate, colon, renal cancers, glioblastoma, and lymphoma.[20–26] Vesigenurtacel-L is the first in type allogeneic whole cell vaccine utilized for the therapy of high-grade, nonmuscle invasive bladder cancer. The vaccine targets patients who do not require immediate radical cystectomy or systemic treatment. The vaccine is made based on using an HSP, which chaperone specific antigens of the tumor to be recognized by the immune system. This is a novel approach for development of vaccine for treatment of bladder cancer, and more experiments should be done to evaluate their efficacy.[22] Allogeneic cancer vaccines are also in experimentation in glioma. In one study, the clinical efficacy of a semi-allogeneic glioma vaccine in mouse model with lethal GL261 gliomas was demostrated. The vaccine was subcutaneously injected to mice and led to protection against glioma in the brain. The tumor-bearing mice were categorized in two groups: one received H-2(b) GL261 glioma cells fused with H-2(d) RAG-neo cells, as semi-allogeneic vaccine and the other received a placebo, i.e. a vaccine of phosphate-buffered saline. The consequence of this strategy showed that the GL261 tumor-conditioned medium increased generation of Th1, inflammatory, and inhibitory cytokines by spleen cells from vaccinated mice with glioma and from control mice. However, these cells and cytokines were decreased in tumor-bearing mice receiving placebo. Altogether, the study revealed a reduce in the growth of the GL261 tumors induced by the vaccine compared to unvaccinated mice with late-stage glioma, which develop splenic dysfunction.[25]

In a study by van Dodewaard-de Jong et al., castration-resistant prostate cancer (CRPC) patients were subjects for studying the efficacy of chemotherapy after immunotherapy with allogeneic cancer vaccine. Since the patients outcome such as improved survival without tumor outcome, i.e. the prolongation of PFS was achieved by the cancer vaccines, this study was conducted to assess the effects of consequent chemotherapy in this regard. In this study, two groups of patients were assessed retrospectively. In the first group, for 23 out of 28 patients who received ipilimumab and GVAX, docetaxel and for 13 out of 28 patients, mitoxantrone was administered. The median PFS for the patients on docetaxel was 6.4 months, whereas the median PFS for patients on mitoxantrone was 4.8 months, which is noticeably increased than what expected. The second group included 21 patients randomized to GVAX or no vaccination. They also received either docetaxel or mitoxantrone. The patients previously treated by GVAX, the median PFS after docetaxel was 9.9 months. The patients who did not receive GVAX, demonstrated a median PSF of 7.1 months after docetaxel (log rank test $p = 0.062$). The median PFS after mitoxantrone was substantially increased in patients who received GVAX as compared with patients not revived the vaccine (5.9 months versus 1.6 months, log rank test $p = 0.0048$). Increased immunologic ratios including CD8+ICOS+ T cell/Treg and pDC/mMDSC ratios, which had correlation with increased PSF in the patients received mitoxantrone, was demostrated. In conclusion, mitoxantrone administered after immunotherapy for CRPC patients, demonstrated improved PFS compared with the studies applied chemotherapy without prior immunohterapy. This might be due to the influence of mitoxantrone on the immune status of patients.[21]

REFERENCES

1. Sondak VK, Sabel MS, Mulé JJ. Allogeneic and autologous melanoma vaccines: where have we been and where are we going? *Clin Canc Res* 2006;**12**(7):2337s–41s.
2. Rivoltini L, et al. Human tumor-derived heat shock protein 96 mediates in vitro activation and in vivo expansion of melanoma-and colon carcinoma-specific T cells. *J Immunol* 2003;**171**(7):3467–74.
3. Lewis JJ. Therapeutic cancer vaccines: using unique antigens. *Proc Natl Acad Sci USA* 2004;**101**(Suppl. 2): 14653–6.
4. Lotem M, et al. Adjuvant autologous melanoma vaccine for macroscopic stage III disease: survival, biomarkers, and improved response to CTLA-4 blockade. *J Immunol Res* 2016;**2016**.
5. Baars A, et al. Skin tests predict survival after autologous tumor cell vaccination in metastatic melanoma: experience in 81 patients. *Ann Oncol* 2000;**11**(8):965–70.
6. Lotem M, et al. Autologous cell vaccine as a post operative adjuvant treatment for high-risk melanoma patients (AJCC stages III and IV). *New Am Joint Committee Cancer Br J Cancer* 2002;**86**(10):1534–9.
7. Convit J, et al. Autologous tumor lysate/Bacillus Calmette–Guérin immunotherapy as an adjuvant to conventional breast cancer therapy. *Clin Transl Oncol* 2015;**17**(11):884–7.
8. Lasalvia-Prisco E, et al. Advanced ovarian cancer: vaccination site draining lymph node as target of immunomodulative adjuvants in autologous cancer vaccine. *Biol Targets Ther* 2007;**1**(2):173.
9. Page DB, et al. Glimpse into the future: harnessing autophagy to promote anti-tumor immunity with the DRibbles vaccine. *J Immunother Cancer* 2016;**4**(1):25.
10. Sanborn RE, et al. A pilot study of an autologous tumor-derived autophagosome vaccine with docetaxel in patients with stage IV non-small cell lung cancer. *J Immunother Cancer* 2017;**5**(1):103.

11. Darefsky AS, King JT, Dubrow R. Adult glioblastoma multiforme survival in the temozolomide era: a population-based analysis of surveillance, epidemiology, and end results registries. *Cancer* 2012;**118**(8): 2163–72.
12. Gilbert MR, et al. A randomized trial of bevacizumab for newly diagnosed glioblastoma. *N Engl J Med* 2014; **370**(8):699–708.
13. Bloch O, et al. Autologous heat shock protein peptide vaccination for newly diagnosed glioblastoma: impact of peripheral PD-L1 expression on response to therapy. *Clin Canc Res* 2017;**23**(14):3575–84.
14. Whiteside TL, et al. Dendritic cell-based autologous tumor vaccines for head and neck squamous cell carcinoma. *Head Neck* 2016;**38**(S1).
15. Katagiri T, et al. Mismatch of minor histocompatibility antigen contributes to a graft-versus-leukemia effect rather than to acute GVHD, resulting in long-term survival after HLA-identical stem cell transplantation in Japan. *Bone Marrow Transplant* 2006;**38**(10):681.
16. Keehn A, Gartrell B, Schoenberg MP. Vesigenurtacel-L (HS-410) in the management of high-grade non-muscle invasive bladder cancer. *Future Oncol* 2016;**12**(23):2673–82.
17. Hsueh EC, Morton DL. Antigen-based immunotherapy of melanoma: Canvaxin therapeutic polyvalent cancer vaccine. In: *Seminars in cancer biology*. Elsevier; 2003.
18. Motl S. Technology evaluation: Canvaxin, John Wayne cancer Institute/CancerVax. *Curr Opin Mol Therapeut* 2004;**6**(1):104–11.
19. Kayaga J, et al. Anti-tumour activity against B16-F10 melanoma with a GM-CSF secreting allogeneic tumour cell vaccine. *Gene Ther* 1999;**6**(8):1475–81.
20. van Dodewaard-de Jong JM, et al. Improved efficacy of mitoxantrone in patients with castration-resistant prostate cancer after vaccination with GM-CSF-transduced allogeneic prostate cancer cells. *OncoImmunology* 2016;**5**(4):e1105431.
21. Keehn A, Gartrell B, Schoenberg MP. Vesigenurtacel-L (HS-410) in the management of high-grade non-muscle invasive bladder cancer. *Future Oncol* 2016;**12**(23):2673–82.
22. Li H, et al. Vaccination with allogeneic GM-CSF gene-modified lung cancer cells: antitumor activity comparing with that induced by autologous vaccine. *Cancer Biother Radiopharm* 2007;**22**(6):790–8.
23. Yu J, Kindy MS, Gattoni-Celli S. Semi-allogeneic vaccine for T-cell lymphoma. *J Transl Med* 2007;**5**(1):39.
24. Gattoni-Celli S, Young MRI. Restoration of immune responsiveness to glioma by vaccination of mice with established brain gliomas with a semi-allogeneic vaccine. *Int J Mol Sci* 2016;**17**(9):1465.
25. Shawler D, et al. Antigenic and immunologic characterization of an allogeneic colon carcinoma vaccine. *Clin Exp Immunol* 2002;**129**(1):99–106.
26. Havranek EG, et al. A novel murine model of allogeneic vaccination against renal cancer. *BJU Int* 2008; **101**(9):1165–9.

CHAPTER

PERSONALIZED CANCER VACCINE

6

Mahsa Keshavarz-Fathi[1,2,3], **Nima Rezaei**[3,4,5]

School of Medicine, Tehran University of Medical Sciences, Tehran, Iran[1]; *Cancer Immunology Project (CIP), Universal Scientific Education and Research Network (USERN), Tehran, Iran*[2]; *Research Center for Immunodeficiencies, Children's Medical Center, Tehran University of Medical Sciences, Tehran, Iran*[3]; *Department of Immunology, School of Medicine, Tehran University of Medical Sciences, Tehran, Iran*[4]; *Network of Immunity in Infection, Malignancy and Autoimmunity (NIIMA), Universal Scientific Education and Research Network (USERN), Tehran, Iran*[5]

INTRODUCTION ON PERSONALIZED MEDICINE

Personalized medicine (PM) is the approach to apply precise treatment to a particular patient, such as like a tailor who makes clothing to fit the individual. In this approach, physicians look for the best clinical response and safety for each patient. Each patient will be diagnosed and treated according to the unique features of their response. Personalized prevention, early diagnoses, risk assessments, applying the best treatment and prognosis could be obtained through application of molecular approaches in medicine to provide the precise strategy for the patient.[1] In this concept, medical interventions are altered according to the genomic, epigenomic, transcriptomic, and proteomic data of each patient, and considering the personal situation of the patient.[2]

PM can decrease the side effects that result due to the trial-and-error approach in medicine and diminish the need to switch to different treatments. It could also decrease the health care costs if applied in the proper setting. The drawbacks of the current conventional approaches, which could be improved by PM, include late diagnosis of disease, severe side effects, and misdiagnosis, which can incur high costs of care. Conventional medical approach can be time-consuming and decrease the patient's satisfaction, which might lead to lower quality of life. Altogether, the objective of PM is to remove the "one-size-fits-all" approach, which is carried out through treatment of the symptoms according to the average clinical response and not the precise response of the patient, especially the high-risk patients.[3,4]

The field of molecular medicine has recently improved sufficiently to develop novel strategies for individualized patient management. New information yielded from the human genome project is able to detect the genetic variations between humans. Approximately 0.1% of the 3 billion bases of human DNA showed variation between individuals. These variations could be used to design clinical trials assessing the patient's response and safety. New technologies for global genomic analysis have been developed such as high-throughput sequencing, single-nucleotide polymorphism genotyping, transcript profiling, and next-generation sequencing (NGS). These new technologies created the medium for using

Vaccines for Cancer Immunotherapy. https://doi.org/10.1016/B978-0-12-814039-0.00006-0

genomic and epigenomic data to develop personalized strategies for each level of health care or in all levels of disease prevention.[5,6]

Various clinical responses to drugs have been observed for centuries. Prediction of these differences as well as adverse effects caused by drugs is not simple. The reasons underlying the differences are now understood from molecular analysis. Variations in genetics and epigenetics are mainly responsible for these differences. Even slight variations and polymorphisms can alter the response of the patient to a drug. In pharmacogenetics, sequence variations in the genes susceptible to influence of response to medication are assessed, however, in pharmacogenomics all of the genes, the genome, are assessed for sequence variations.[7] Pharmacodiagnostics or theranostics means utilizing molecular analysis to apply the best modality for diagnosis, prevention, and treatment for an individual.[8]

PERSONALIZED MEDICINE IN CANCER PATIENTS

Cancer is one of the targets of personalized medicine because a particular cancer can be heterogeneous between patients, in term of pathogenesis, susceptibility to metastasis, immunologic responses, and clinical responses to treatment.[9,10] Underlying genetic and epigenetic alterations, which are the basis of PM, play a role in the pathogenesis of cancer. PM can be applied in screening, prevention, diagnosis, prognosis, prediction of clinical response to therapy, early detection of metastasis or disease recurrence, and the categorizing of patients into particular subgroups who receive personalized treatment.[5,11]

SCREENING OF CANCER

Both genetic and environmental determinants have been proposed as cancer risk factors. Identifying the contributors precisely can lead to different approaches for prevention or early diagnosis of cancer.[12] Modification to life style might be utilized in case of environmental risk factors.[13] However, in case of genetic variation, preventive interventions such as biological or surgical approaches, are recommended. For example, in the people with mutations in breast cancer susceptibility gene 1 (BRCA1) or 2 (BRCA2), prophylactic resection of breasts, frequent screenings, or adjuvant therapies are performed. These genes can also induce malignancy in other organs such as ovary and prostate, therefore, prophylactic salpingo-oophorectomy might be carried out as well.[14–18]

CLASSIFICATION OF TUMORS

Classifying tumors by molecular analysis including DNA, RNA, microRNA (miRNA), or protein rather than the traditional classification by only histologic features comes with a lot of advantages. This could provide useful information on the individualized treatment candidates, the prognosis of the patient, and susceptibility to recurrence or metastasis.[19]

TARGETED THERAPY AND USING PREDICTIVE BIOMARKERS

Recognizing the responders and nonresponders to a particular therapy can be possible through understanding the molecular basis of the disease, which might vary from patient to patient. These molecular variations and differences in response to treatment can be the basis for development of new medications. Even patients with the same clinical presentation and histologic type of tumor might show various molecular profiles, which would affect their response to treatment. Alternative therapeutic options must be considered for patients, who will not respond to a particular therapy. The molecular variations in nucleic acids and proteins between subgroups of patients can be applied as predictive biomarkers to select or design the appropriate therapies.[20,21]

Predictive biomarkers are also utilized for immunotherapeutic modalities. The genes such as CD3, CD40, CD27, CD38, CD8A, GZMB, PRF1, and CCL4, which play a role in immune functions, are among the predictive biomarkers. Higher expression of these genes was associated with better response to treatment in patients with melanoma taking ipilimumab.[22]

SAFETY OF MEDICATION

The dose of medication or its safety is dependent on the molecular alterations as well. This can be predicted through genetic and genomic variations. Sequence variations in genes, which are involved in pharmacokinetics of drugs or encode the target of the treatment, influence the safety and clinical outcomes of medications. Polymorphisms in cytochrome P450 (CYP) enzymes, which are involved in the metabolism of drugs, are a known instance of the effects of genetic variation on metabolism and safety of drugs.[21,23]

PROGNOSTIC BIOMARKERS

Decisions for patients with cancer are typically made based on the prognostic factors i.e., clinicopathologic characteristics of the patient including the histological features, grade and stage of the cancer, and limited biochemical tumor markers. Nevertheless, the molecular profile of the patient has evolved into the prognostic model of cancer. The combination prognostic model, using both clinicopathologic and molecular parameters, has surpassed the conventional method.[24–26] For instance, T cell responses are associated with improvement of clinical outcome. There is evidence of prognosticating overall survival and progression-free survival due to spatial distribution of lymphocytes and their infiltration into the tumor. Two subclasses of lymphocytes demonstrated different patterns of infiltration into the tumor and consequently different survival outcomes. According to the immunological biomarkers, an immunescore is considered for anticipation of patients' survival.[27,28]

NEOANTIGENS AND PERSONALIZED CANCER VACCINES

The field of immunopharmacogenomics has developed due to advances in genomic technologies. It aims to apply the genomic data of tumor cells to design individualized immunotherapies. Neoantigens, which are unique and tumor-specific antigens, are attractive targets for specific immunotherapy. Adverse effects of treatment are minimized in this tumor-specific approach. New technologies should be utilized for prediction, identification, and validation of neoantigens.[29]

Somatic mutations in cancerous cells are responsible for expression of neoantigens, which are not expressed by the normal cells and consequently targeting these antigens does not lead to toxicity. Different types of vaccines including dendritic cell, synthetic long peptide, and RNA-targeting neoantigens have resulted in both CD4+ and CD8+ T cell responses, which are specific to the neoantigens.[30–32]

NGS technologies are suitable for rapid and massive analysis of the genome to compare the normal cells and cancer cells in terms of somatic mutations. The majority of these mutations might not lead to gene expression. However, the expressed mutation results in production of an abnormal peptide, which is presented by major histocompatibility complex (MHC) molecules to the immune cells. There are several considerations necessary to select the neoantigen proper for vaccination. The characteristics of a candidate neoantigen include the following items: the somatic mutation must be expressed by the tumor cell, the neoantigen must be processed and presented by MHC molecules, it must have high-affinity to MHC, and the MHC/neoantigen complex must have high affinity to the T cell receptor.[30,33–35]

Since a lot of somatic mutations are usually found in the tumor cells, and due to the polymorphism of human MHC molecules, it is not simple to select epitopes for inclusion in vaccine. Therefore, different methods have been developed for prediction and prioritization of epitopes. *In silico* approaches such as sequence-based and structure-based algorithms have been developed to assist in this step. Afterwards, immunogenicity of the neoantigens is validated through *in vitro* and *in vivo* tests. Following these steps, the neoantigen vaccine is evaluated in clinical trials.[30,36–38]

PERSONALIZED CANCER VACCINES IN CLINICAL STUDIES

WHOLE TUMOR CELL VACCINES

Whole tumor cells are among the first personalized cancer vaccines. In this vaccine type, all the relevant antigens are available in the tumor cells and identification of the appropriate and immunogenic antigen is not required. The efficacy of this vaccine is improved by engineering to secrete cytokines such as GM-CSF. Other types of cancer vaccines have the potential to be designed as individualized vaccines.[39] In the following discussion, we will review the available personalized vaccines from different categories.

OncoVAX (Vaccinogen, Inc.) is a vaccine administered to patients with colorectal cancer after surgical removal, aiming to extend patients' survival and decrease the rate of recurrence. It is an autologous whole tumor cell vaccine administered with or without the adjuvant Bacillus Calmette-Guerin (BCG). It showed improvement of survival in stage 2 colorectal cancer.[40]

Autologous tumor cells bound to the hapten, dinitrophenyl (DNP) can increase the immunogenicity of vaccine and amplify the immune responses. This approach is utilized for preparing M-Vax (AVAX, Inc.) for melanoma or O-vax for ovarian cancer. BCG is administered as vaccine adjuvant as well. In a study, M-Vax was administered to stage 3 melanoma patients and led to 19%–24% increase in 5-year survival rate compared to surgery alone.[41]

Melacine, which is an autologous tumor cell vaccine administered to patients with melanoma, includes the lysate of two melanoma cell lines, MSM-M-1 and MSM-M-2, with Detox-PC as adjuvant. Detox-PC consists of monophosphoryl lipid A (MPL) and mycobacterial cell wall skeleton, and the cell lines express different melanoma antigens, including gp100, MAGE-1–3, Melan-A/MART, tyrosinase, TRP-1, and TRP-2. The vaccine was approved in Canada based on the results of a phase 3 trial comparing Melacine with chemotherapy.[39,42,43]

Oncophage (vitespen, Antigenics) is an individualized cancer vaccine consisting of the heat shock protein gp96 containing the tumor antigens released form autologous tumor cells. Processing the vaccine is obtained from removal of a sample from the tumor, and it is administered via intradermal injection to the patients after surgical resection of the tumor. The vaccine has been evaluated in various types of cancer and it obtained approval in Russia as adjuvant therapy for renal cell carcinoma at intermediate risk of recurrence. Oncophage has orphan drug designation from the US Food and Drug Administration (FDA) and the European Medicines Agency for renal carcinoma and glioma.[39] It has been administered for various tumor types in clinical trials such as melanoma, renal cell carcinoma, gastric, pancreatic, colorectal, and ovarian cancers, and lymphoma and chronic myelogenous leukemia.[44]

It was evaluated in patients with stage 3 or stage 4 metastatic melanoma as well. It was deemed safe and capable of inducing both innate and tumor-specific immune response and was more effective in the early stages. The results were remarkable because even tiny samples could provide a sufficient source for preparing the vaccine.[45,46]

In the phase 3 study addressing efficacy and safety of vitespen, 319 patients with stage 4 melanoma randomly received the vaccine or physician's choice (PC) of a treatment. PC included various conventional therapies such as complete resection with or without other therapies, chemotherapy (dacarbazine or temozolomide), or interleukin 2. The 133 patients in the vitespen group and 86 patients in the PC group were analyzed. Significant adverse effects were not seen in vaccine group except for one cellulitis and one thyroid function disorder, which resolved by treatment. Intention-to-treat analysis of survival was done for all patients and stratified by substages according to the American Joint Committee on Cancer, which ended up with no significant differences. A secondary analysis was done comparing vaccine and PC groups stratified by the number of interventions (more than 1 or 10 and more than 10 treatment doses). A trend in favor of vitespen was observed for those receiving 10 or more than 10 doses of vaccine but was not statistically significant. However, a significant difference in survival was seen in the substages of M1a plus M1b who received 10 or more than 10 vitespen doses.[46]

PEPTIDE/PROTEIN-BASED VACCINE

BIOVAXID (Accentia BioPharmaceuticals) is one of the idiotype (Id) vaccines evaluated for B cell lymphoma, mainly follicular lymphoma. In this approach, normal B cells from the patient's lymph nodes are used to find the personalized Id protein, a variable region of immunoglobulins on the surface of B cells. The protein is administered to the patient with an adjuvant such as keyhole limpet hemocyanin (KLH). This vaccine is promising immunotherapy for treatment of follicular lymphoma, which frequently relapses after surgery. It has been evaluated in a phase 3 trial for increasing the relapse-free survival of patients with follicular lymphoma. The median time to relapse for the treatment group was 44.2 months, versus 30.6 months for the control group ($P = 0.045$; HR $= 1.6$).[47,48] MyVax (Genitope Corp.) is the other Id protein-based vaccine evaluated for treatment of follicular non-Hodgkin lymphoma.[49]

Recent advances in genomic studies have paved the way for personalized vaccine development. The neoantigens identified through new technologies are promising candidates for treatment of cancers at high risk of recurrence. For instance, in the trial by Ott et al. in six patients with stage 3 or 4 melanoma, who had undergone resection surgery and were at high risk of recurrence, vaccination with up to 20 individualized neoantigens were done. They took samples from the tumor and peripheral blood of each patient and performed whole-exome sequencing to identify somatic mutations. The expression of mutated alleles were also examined through RNA-sequencing and the neoantigen was assessed in terms of binding and affinity to HLA-A and -B. Afterward, the long peptides containing up to 20 top ranked individualized neoantigens plus poly-ICLC, the TLR3 agonist, as adjuvant were administered to the patients. Twenty-five months after vaccination, 4 out of 6 patients did not have recurrent disease. Although two patients had recurrence, treatment with immune checkpoint blockade, antiprogrammed cell death-1 (anti PD-1) antibody, led to complete regression of tumor.[2,50]

IMMUNE CELL–BASED VACCINES

Provenge (Sipuleucel-T, Dendreon Corp.) is an autologous immune cell–based vaccine. Autologous peripheral-blood mononuclear cells are extracted, which consist of antigen presenting cells, and are incubated *ex vivo* with prostatic acid phosphatase (PAP) and GM-CSF. The vaccine has shown improved survival compared to placebo (25.8 vs. 21.7 months) in a phase 3 study in men with castration-resistant prostate cancer, which led to its FDA approval. Provenge is the first therapeutic cancer vaccine approved by FDA.[51,52]

DCVax (Northwest Biotherapeutics) uses an individualized approach of extracting dendritic cells (DCs) from peripheral blood monocytes of the patients. The DCs are loaded with autologous tumor antigens to cross present the tumor antigen to T cells and induce a specific immune response. Safety, immune responses, and clinical responses have been yielded in the study in patients with high-grade glioma. DCVax-Brain has received orphan drug designation from FDA.[53,54]

Imetelstat (Geron Corp.'s GRNVAC1) is composed of autologous DCs pulsed *ex vivo* with telomerase reverse transcriptase (hTERT) mRNA. hTERT is overexpressed in many types of tumor and is a good candidate for cancer immunotherapy. Imetelstat has been evaluated for metastatic prostate cancer, and showed safety and increase in prostate specific antigen (PSA) doubling time.[55,56]

Autologous DCs are genetically engineered *ex vivo* to create a sustained source of tumor antigens. Recombinant nonreplicating virosomes are used to carry and express a particular gene to DCs. Although virosomes act as adjuvants to augment the immune response developed by the vaccine, those that produce viral antigens might result in immune responses against the cells carrying them, i.e., the DCs. Utilization of virosomes such as retroviral or adenoviral vectors does not lead to production of viral antigens. Lentiviral vectors are the other candidates, with an advantage over retroviruses. They are capable of genetic induction without replication of the host.[57,58]

DCs, which are fused with the autologous tumor cells, are suitable vaccines, which consist of both tumor antigens and antigen presenting cells (APCs). This is the combination of DC-based and autologous whole tumor cell vaccines, which make a personalized approach for vaccination.[59]

GENETIC-BASED VACCINE

Neoantigen identification is useful for designing a personalized genetic-based vaccine. RNA-based poly-neo-epitope approach is applied to induce immune responses against the individualized neoantigen, to which a strong immune response might not exist before vaccination.[60] This approach was recently examined in 13 patients with stages 3 and 4 melanoma. Neoantigens, identified via exome sequencing and RNA sequencing, and those with high affinity to MHC class I and II were selected as immunogenic neoantigens for immunization. The patients were treated with poly-neo-epitope RNA vaccines and all of them showed increased progression-free survival. Eight out of 13 patients who had no lesion prior to vaccination, remained completely recurrence-free for 12–23 months. Among the five patients with lesions, the results included two complete responses, one partial response, and one stable disease. One of the patients showed complete response by combination of vaccine with PD-1 blockade.[61]

REFERENCES

1. Vogenberg FR, Isaacson Barash C, Pursel M. Personalized medicine: part 1: evolution and development into theranostics. *Pharm Ther* 2010;**35**(10):560–76.
2. Bullen Love D. The potential of personalized cancer vaccines. *Oncol Times* 2017;**39**(18):1,8–8.
3. Abrahams E, Ginsburg GS, Silver M. The personalized medicine coalition: goals and strategies. *Am J Pharmacogenomics* 2005;**5**(6):345–55.
4. Mathur S, Sutton J. Personalized medicine could transform healthcare. *Biomed Rep* 2017;**7**(1):3–5.
5. Diamandis M, White NMA, Yousef GM. Personalized medicine: marking a new epoch in cancer patient management. *Mol Cancer Res* 2010;**8**(9):1175–87.
6. Rabbani B, et al. Next generation sequencing: implications in personalized medicine and pharmacogenomics. *Mol Biosyst* 2016;**12**(6):1818–30.
7. Mancinelli L, Cronin M, Sadee W. Pharmacogenomics: the promise of personalized medicine. *AAPS PharmSci* 2000;**2**(1):E4.
8. Jeelani S, et al. Theranostics: a treasured tailor for tomorrow. *J Pharm BioAllied Sci* 2014;**6**(Suppl. 1):S6–8.
9. Fidler IJ. Tumor heterogeneity and the biology of cancer invasion and metastasis. *Cancer Res* 1978;**38**(9): 2651–60.
10. Whiteside TL. Immune responses to cancer: are they potential biomarkers of prognosis? *Front Oncol* 2013;**3**: 107.
11. Verma M. Personalized medicine and cancer. *J Personalized Med* 2012;**2**(1):1–14.
12. Perera FP. Environment and cancer: who are susceptible? *Science* 1997;**278**(5340):1068–73.
13. Overdevest JB, Theodorescu D, Lee JK. Utilizing the molecular gateway: the path to personalized cancer management. *Clin Chem* 2009;**55**(4):684–97.
14. Meijers-Heijboer H, et al. Use of genetic testing and prophylactic mastectomy and oophorectomy in women with breast or ovarian cancer from families with a BRCA1 or BRCA2 mutation. *J Clin Oncol* 2003;**21**(9): 1675–81.
15. Grann VR, et al. Decision analysis of prophylactic mastectomy and oophorectomy in BRCA1-positive or BRCA2-positive patients. *J Clin Oncol* 1998;**16**(3):979–85.
16. Gronwald J, et al. Tamoxifen and contralateral breast cancer in BRCA1 and BRCA2 carriers: an update. *Int J Cancer* 2006;**118**(9):2281–4.
17. Cavanagh H, Rogers KMA. The role of BRCA1 and BRCA2 mutations in prostate, pancreatic and stomach cancers. *Hered Cancer Clin Pract* 2015;**13**:16.
18. Kauff ND, et al. Risk-reducing salpingo-oophorectomy for the prevention of BRCA1- and BRCA2-associated breast and gynecologic cancer: a multicenter, prospective study. *J Clin Oncol* 2008;**26**(8):1331–7.
19. De Mattos-Arruda L, Rodon J. Pilot studies for personalized cancer medicine: focusing on the patient for treatment selection. *Oncologist* 2013;**18**(11):1180–8.
20. Allison M. Is personalized medicine finally arriving? *Nat Biotechnol* 2008;**26**(5):509–17.
21. Hayes DF, et al. Personalized medicine: risk prediction, targeted therapies and mobile health technology. *BMC Med* 2014;**12**:37.
22. Fridman WH, et al. The immune contexture in human tumours: impact on clinical outcome. *Nat Rev Cancer* 2012;**12**(4):298–306.
23. Mlecnik B, et al. Integrative analyses of colorectal cancer show immunoscore is a stronger predictor of patient survival than microsatellite instability. *Immunity* 2016;**44**(3):698–711.
24. Ji RR, et al. An immune-active tumor microenvironment favors clinical response to ipilimumab. *Cancer Immunol Immunother* 2012;**61**(7):1019–31.
25. Chen Q, Wei D. Human cytochrome P450 and personalized medicine. *Adv Exp Med Biol* 2015;**827**:341–51.
26. Kulasingam V, Diamandis EP. Strategies for discovering novel cancer biomarkers through utilization of emerging technologies. *Nat Clin Pract Oncol* 2008;**5**(10):588–99.

27. Kim HJ, et al. Virtual-karyotyping with SNP microarrays in morphologically challenging renal cell neoplasms: a practical and useful diagnostic modality. *Am J Surg Pathol* 2009;**33**(9):1276–86.
28. Hanbali A, et al. The evolution of prognostic factors in multiple myeloma. *Adv Hematol* 2017;**2017**:4812637.
29. Kiyotani K, Chan HT, Nakamura Y. Immunopharmacogenomics towards personalized cancer immunotherapy targeting neoantigens. *Cancer Sci* 2018;**109**(3):542–9.
30. Li L, Goedegebuure SP, Gillanders WE. Preclinical and clinical development of neoantigen vaccines. *Ann Oncol* 2017;**28**(Suppl. 12):xii11–7.
31. Harjes U. Tumour vaccines: personal training by vaccination. *Nat Rev Cancer* 2017;**17**(8):451.
32. Varypataki EM, et al. Synthetic long peptide-based vaccine formulations for induction of cell mediated immunity: a comparative study of cationic liposomes and PLGA nanoparticles. *J Control Release* 2016;**226**: 98–106.
33. Efremova M, et al. Neoantigens generated by individual mutations and their role in cancer immunity and immunotherapy. *Front Immunol* 2017;**8**:1679.
34. Lu Y-C, Robbins PF. Cancer immunotherapy targeting neoantigens. *Semin Immunol* 2016;**28**(1):22–7.
35. Schumacher TN, Schreiber RD. Neoantigens in cancer immunotherapy. *Science* 2015;**348**(6230):69–74.
36. Rane SS, Javad JMS, Rees RC. Tumor antigen and epitope identification for preclinical and clinical evaluation. In: Rezaei N, editor. *Cancer immunology: bench to bedside immunotherapy of cancers*. Berlin, Heidelberg: Springer Berlin Heidelberg; 2015. p. 55–71.
37. Karasaki T, et al. Prediction and prioritization of neoantigens: integration of RNA sequencing data with whole-exome sequencing. *Cancer Sci* 2017;**108**(2):170–7.
38. March-Vila E, et al. On the integration of in silico drug design methods for drug repurposing. *Front Pharmacol* 2017;**8**:298.
39. Jain KK. Personalized cancer vaccines. *Expet Opin Biol Ther* 2010;**10**(12):1637–47.
40. Hanna JMG. Immunotherapy with autologous tumor cell vaccines for treatment of occult disease in early stage colon cancer. *Hum Vaccines Immunother* 2012;**8**(8):1156–60.
41. Berd D. M-Vax: an autologous, hapten-modified vaccine for human cancer. *Expert Rev Vaccines* 2004;**3**(5): 521–7.
42. Fintor L. Melanoma vaccine momentum spurs interest, investment. *J Natl Cancer Inst* 2000;**92**(15):1205–7.
43. Dubensky TW, Reed SG. Adjuvants for cancer vaccines. *Semin Immunol* 2010;**22**(3):155–61.
44. di Pietro A, et al. Oncophage: step to the future for vaccine therapy in melanoma. *Expert Opin Biol Ther* 2008;**8**(12):1973–84.
45. Eton O, et al. Autologous tumor-derived heat-shock protein peptide complex-96 (HSPPC-96) in patients with metastatic melanoma. *J Transl Med* 2010;**8**:9.
46. Testori A, et al. Phase III comparison of vitespen, an autologous tumor-derived heat shock protein gp96 peptide complex vaccine, with physician's choice of treatment for stage IV melanoma: the C-100-21 Study Group. *J Clin Oncol* 2008;**26**(6):955–62.
47. Mahaseth H, et al. Idiotype vaccine strategies for treatment of follicular lymphoma. *Future Oncol* 2011;**7**(1): 111–22.
48. Schuster SJ, et al. Idiotype vaccine therapy (BiovaxID) in follicular lymphoma in first complete remission: phase III clinical trial results. *J Clin Oncol* 2009;**27**(18S):2.
49. Timmerman JM, Vose JM, Czerwinski DK. *Leuk Lymphoma* 2009;**50**(null):37.
50. Ott PA, et al. An immunogenic personal neoantigen vaccine for patients with melanoma. *Nature* 2017; **547**(7662):217–21.
51. Kantoff PW, et al. Sipuleucel-t immunotherapy for castration-resistant prostate cancer. *N Engl J Med* 2010; **363**(5):411–22.
52. Yousefi H, et al. Immunotherapy of cancers comes of age. *Expert Rev Clin Immunol* 2017;**13**(10):1001–15.

53. Polyzoidis S, Ashkan K. DCVax®-L—developed by Northwest Biotherapeutics. *Hum Vaccines Immunother* 2014;**10**(11):3139–45.
54. Wheeler CJ, Black KL. *Expert Opin Investig Drugs* 2009;**18**(null):509.
55. Su Z, et al. Telomerase mRNA-Transfected dendritic cells stimulate antigen-specific $CD8^+$ and $CD4^+$ T cell responses in patients with metastatic prostate cancer. *J Immunol* 2005;**174**(6):3798–807.
56. Roth A, Harley CB, Baerlocher GM. Imetelstat (GRN163L)–telomerase-based cancer therapy. *Recent Results Cancer Res* 2010;**184**:221–34.
57. Pincha M, et al. Lentiviral vectors for induction of self-differentiation and conditional ablation of dendritic cells. *Gene Ther* 2011;**18**(8):750–64.
58. Bubenik J. Genetically engineered dendritic cell-based cancer vaccines (review). *Int J Oncol* 2001;**18**(3): 475–8.
59. Koido S. Dendritic-tumor fusion cell-based cancer vaccines. *Int J Mol Sci* 2016;**17**(6):828.
60. Hellmann MD, Snyder A. Making it personal: neoantigen vaccines in metastatic melanoma. *Immunity* 2017; **47**(2):221–3.
61. Sahin U, et al. Personalized RNA mutanome vaccines mobilize poly-specific therapeutic immunity against cancer. *Nature* 2017;**547**(7662):222–6.

CHAPTER

WHOLE TUMOR CELL VACCINE FOR CANCER

7

Sepideh Razi[1,2], Mahsa Keshavarz-Fathi[1,3,4]

Cancer Immunology Project (CIP), Universal Scientific Education and Research Network (USERN), Tehran, Iran[1]; Student Research Committee, School of Medicine, Iran University of Medical Sciences, Tehran, Iran[2]; School of Medicine, Tehran University of Medical Sciences, Tehran, Iran[3]; Research Center for Immunodeficiencies, Children's Medical Center, Tehran University of Medical Sciences, Tehran, Iran[4]

Cancer vaccines with defined antigens have been used to immunize patients against tumors for several years. However, these vaccines have several disadvantages. Immunization of patients is limited to the antigens, which were used in the vaccine, while there are still several tumor antigens, which are uncharacterized. Additionally, tumor cells undergo mutations, which could result in loss of antigens. Therefore, whole tumor cell vaccines consisting of allogeneic or autologous tumor cells or tumor cell lines were developed. Whole tumor cell vaccines comprise all of the defined and unidentified tumor antigens. Antigen presenting cells (APCs) present these tumor-associated antigens (TAAs) by major histocompatibility complex (MHC) class I and II to CD8+ cytotoxic T lymphocytes (CTLs) and CD4+ T cells, respectively, leading to activation of immune responses against tumor. CTLs can lead to apoptosis by interaction of Fas/Fas ligand or by secretion of lytic enzymes. Activation of CD4+ T cells stimulates the activity of natural killer (NK) cells. In addition, these cells activate humoral immune response and could also promote the stimulation of CD8+ T cells.[1,2] Different methods have been used to prepare whole tumor cell vaccines such as using tumor cell lysates, tumor-derived exosomes, and irradiated gene-modified tumor cell lines.

TUMOR CELL LYSATES

Whole tumor cell lysates contain tumor cell antigens. Autologous or allogeneic lysates could be utilized for cancer vaccination. There are several methods to prepare tumor lysates such as being frozen and thawed repeatedly[3] or using ultraviolet B (UVB) radiation[4] leading to tumor cell death. Necrotic tumor cells release several proteins such as high mobility group box 1 (HMGB1)[5] and heat shock proteins (HSPs) as well as DNA and RNA, which all are among danger signals. It has been shown that tumor cell lysates can trigger DC maturation. Different pathways have been suggested to explain this effect, including HMGB1/RAGE pathway, which is released by necrotic cells and leads to activation of proinflammatory functions and DCs maturation.[6] Release of HSPs leads to increased levels of costimulatory molecules and DC markers including CD40, CD83, and CD86; and consequently DCs maturation.[7] In addition, breakdown of DNA and RNA produces uric acid, which could lead to DCs activation.[8]

Vaccines for Cancer Immunotherapy. https://doi.org/10.1016/B978-0-12-814039-0.00007-2

TUMOR-DERIVED EXOSOMES

Exosomes are membrane vesicles that are produced by multivesicular endosomes of cells in both physiological and pathological conditions and are secreted into the extracellular matrix via exocytosis. It has been shown that tumor cell–derived exosomes could be obtained from ascites, pleural effusion, and plasma samples of patients. Exosomes comprise tumor antigens such as mesothelin, CEA, EGFR2, HER-2/neu, TRP-1, MART-1, gp100 and heat-shock protein 70 (HSP 70) and HSP 90.[9–13] HSPs are highly conserved proteins that are expressed in all cells, and their level of expression is increased in response to cellular stress.[14] These molecules have the ability to stabilize and chaperone peptides and proteins and they can bind to different peptides of proteins that exist in the cells such as normal proteins that are expressed in the cells or mutated proteins that are associated with cancer. When the cell membrane ruptures due to a cell damage or death, these HSP-peptide complexes are released to the extracellular space.[15] These complexes can interact with APCs such as DCs via HSP receptor called CD91 in extracellular space, and then DCs present these peptides to immune system by MHC class I. Therefore, HSPs can trigger the activation of immune responses against tumor by carrying antigenic peptides. In addition, HSPs have the ability to stimulate innate immunity such as NK cells.[16] Therefore, based on these characteristics of HSPs, these molecules were used for developing cancer vaccines.

IRRADIATED GENE-MODIFIED TUMOR CELL VACCINE

Irradiated gene-modified tumor cell line vaccines consisting of autologous or allogenic tumor cells were genetically engineered to encode immunostimulatory agents[17] including cytokines such as tumor necrosis factor-α,[18] granulocyte-macrophage colony-stimulating factor (GM-CSF)[19] and IL-4,[20] and costimulatory molecules such as CD80.[21] Local cytokines trigger the activation of NK cells and T cells. Also, these cytokines could stimulate antitumor inflammatory responses.[22]

CLINICAL TRIALS

Belagenpumatucel-L (Lucanix), which is a tumor cell vaccine derived from transforming growth factor-beta2 (TGF-β2) antisense gene-modified whole non-small cell lung cancer (NSCLC) cell lines, was tested in a randomized phase II study of 75 patients with stages II, III, and IV NSCLC. Patients were randomly divided into three groups and were given up to 16 administration of one of the doses of intradermal Belagenpumatucel-L, i.e. 1.25, 2.5 or 5×10^7 cells on a monthly or every other month treatment cycle. Among 61 patients with advanced stage of NSCLC, partial response was 15%. Participants who received the highest dose of vaccine had a noticeable overall survival (OS) advantage compared to the patients who received the low dose ($P = 0.0069$). The probabilities of 1- and 2-year survival were estimated as well, which were as follows: 68% and 52%, respectively, for the patients in both high dose arms compared with 39% and 20% for patients in the lowest dose arm. An increase in the blood level of cytokines was also observed in patients who responded to Belagenpumatucel-L (IL-6, $P = .004$; IFN-γ, $P = .006$; and IL-4, $P = 0.007$). Additionally, these patients had an increased humeral immune response. An improvement on the OS was also observed in the patients responded to the vaccine compared with those did not ($P = 0.011$). In this trial, one injection site edema and 16% flu-like symptoms were observed. Therefore, this vaccine showed no serious adverse effects.[23] Another phase III study was conducted for this vaccine. A total of 532 NSCLC patients with stage IIIA, IIIB, and IV who did not show any progression after platinum-based chemotherapy were enrolled in this study. Patients were randomly divided into two groups; 270 patients were given Belagenpumatucel-L and 262 participants received placebo. Patients received 18 doses of Belagenpumatucel-L intradermally in a monthly cycle of treatment and two booster doses at months 21 and 24. The OS of the participants who received vaccine was 20.3 versus 17.8 months for patients who received placebo ($P = 0.594$), which was not significant. Subgroup analysis showed the OS of patients who received vaccine within 12 weeks of completing platinum-based chemotherapy was noticeable. In this subgroup, the OS of participants who were given vaccine was 20.7 versus 13.4 months for patients who received placebo ($P = 0.083$). Additionally, improved OS was observed in patients who did not have adenocarcinoma, 19.9 versus 12.3 months for placebo ($P = 0.036$), and in patients who received radiotherapy during their chemotherapy, 40.1 versus 10.3 months for placebo ($P = 0.014$). This vaccine only showed injection site reactions and did not have any important side effects.[24]

GVAX is a genetically modified vaccine based on the using irradiated tumor cells, as whole tumor cells, which are also engineered to release the cytokine GM-CSF. In 2015, for the first time, a clinical study was conducted to test the effect and safety of GVAX in patients with melanoma. After resection of tumor, a total of 20 patients with stage IIB-IV melanoma were enrolled in this study. Patients received 4 doses of this vaccine intradermally in 28-day cycles in combination with cyclophosphamide, which shows immunomodulatory function through decreasing the population of regulatory T cells (Tregs). Eighteen out of twenty patients completed four cycles of vaccination. The results showed the increased serum level of GM-CSF in a dose-dependent manner. In addition, elevated numbers of activated monocytes were detected in the circulation. No significant activation of specific immune responses against melanoma tumor was observed. Elevated levels of PD-1+ lymphocytes and eosinophils were reported at the injection site of the vaccine. Cyclophosphamide, which was administered in combination with GVAX, did not decrease the numbers of Tregs in the circulation. No significant side effects were reported in this study. Only vitiligo lesion and white hair patches were observed in one patient.[25]

A phase I clinical study has tested GVAX in patients with stage IV NSCLC. This study showed 18 out of 25 patients showed increased infiltration of granulocytes, lymphocytes, DCs, and macrophages and, additionally, improved remissions were demonstrated in patients who received this vaccine. Adverse effects were limited to grade 1 and 2 local skin reactions.[26] Another phase I/II study was conducted in individuals with stage IB or II NSCLC and patients with stage III or IV disease. It showed better outcomes in patients who received vaccine releasing higher GM-CSF. Therefore, this trial represented a correlation between the secretion of GM-CSF and patient's survival.[27] This vaccine has also been utilized in prostate cancer. However, it did not have promising results in phase II clinical trials.[28,29]

In 1997, a phase I clinical trial investigated an autologous GM-CSF gene-modified tumor vaccine (GMTV) in patients with metastatic renal cell carcinoma (mRCC). This trial did not show any noticeable results in clinical outcome. However, it showed increased tendency toward delayed-type hypersensitivity (DTH).[30] Five years later, another phase I study for GMTV was conducted using CD80 gene in conjunction with IL-2. The results of the total 13 patients showed 2 patients with partial response and 2 patients with stable disease out of 13. Also, 3 out of 4 of these participants developed DTH.[21]

A clinical study tested the effect of a whole tumor cell vaccine in patients with advanced chronic lymphocytic leukemia (CLL). This vaccine consisted of irradiated tumor cells in combination with bystander cells, which secrete GM-CSF. Patients received this whole tumor cell vaccine after allogeneic hematopoietic stem cell transplantation. The results showed the 2-year progression free survival (PFS) was 82% and the OS was 88% after 2.9 years of follow-up. In addition, activation of CD8+ T cells against tumor cells were observed in the patients who received this vaccine after transplantation. Therefore, this trial suggested that administration of this vaccine to patients with CLL after transplantation can be helpful in controlling the disease.[31]

Another phase I clinical study was conducted to test the effect of a whole tumor cell vaccine in patients with grade 4 astrocytoma. In this vaccine the expression of TGF-β2, which is an immunosuppressive agent, was blocked. A total of 6 patients were given 2–7 subcutaneous doses of this vaccine, which consisted of genetically modified autologous tumor cells. The results showed 53%–98% blockade of the release of TGF-β2 by cancer cells. Two out of six patients showed SD and two others showed partial regression. In addition, the results reported that immune responses were activated by this vaccine. The median OS of all patients enrolled in the study was 68 weeks, while the OS of responding patients was 78 weeks. This vaccine was well tolerated and no serious adverse effects were observed.[32]

In order to determine the efficacy of a whole tumor cell vaccine, a clinical study was conducted in patients with melanoma. Patients who underwent the removal surgery of tumor greater than or equal to 5 g were enrolled in this study and received injections of UVB-irradiated autologous tumor cells intradermally in combination with DETOX as an adjuvant. The immune responses of 35 patients were analyzed. Ten out of 35 patients showed the induction of antitumor peripheral blood mononuclear cell (PBMC) cytotoxicity. DTH responses to tumor cells were reported in 8 patients. This vaccine was well tolerated and no serious side effects were observed.[33]

Melacine contains tumor cell lysates from two melanoma cell lines and Detox adjuvant.[34] In a phase III clinical trial this vaccine was assessed for stage II melanoma patients. The outcomes in favor of the vaccine group, were not significant. However, analysis showed that the OS is increased in patients who had at least two out of five HLA antigens of this vaccine, especially HLA-A2 and C3.[35]

A randomized phase III clinical trial was conducted in patients with stage III renal cancer after nephrectomy for Retionale, which is a tumor cell lysate vaccine. This trial showed increased 5-year progression-free survival (PFS).[36]

A phase I clinical trial was conducted to test an exosome vaccine. Forty patients were randomly divided into two groups. One group received the vaccine, i.e. subcutaneous ascites-derived exosomes, which express CEA, and the other group was given subcutaneous ascites-derived exosomes that express CEA in combination with GM-CSF, weekly for 4 months. The results showed that 75% of the participants who received exosomes with GM-CSF showed activated CTL response, whereas the activation of CTL response was observed only in 20% of the patients of the other group. This trial suggested that exosome in combination with GM-CSF could be used as immunotherapy of colorectal cancer patients. The adverse effects were mild, including local injection reactions (grades 1 and 2).[37]

APPROVED VACCINE

Melacine is a tumor vaccine for treatment of melanoma. This vaccine consists of whole tumor cell lysates. Detoxified bacterial endotoxin adjuvant, which is composed of lipopolysaccharides and mycobacterial cell-wall skeleton, is also included in this vaccine. Melacine was approved by the Canadian Health Protection Branch in 1999 after a phase III study in patients with stage IV melanoma. The results of this trial did not show a significant survival advantage for the vaccine, however, a remarkable improvement of quality of life of patients who received Melacine versus patients who underwent chemotherapy led to this license.[38]

A study compared the OS of patients with melanoma who received Melacine in combination with low-dose interferon alpha versus patients who were given high-dose interferon alpha. However, the results demonstrated no significant differences between these two groups.[39] Same outcomes were observed in a phase III clinical study. Patients were randomly divided in two groups in this study, and one group received Melacine. Five-year estimated disease-free survival for patients who received this vaccine was 65% versus 63% for the other group.[40] Therefore, these two studies did not show a survival benefit of this vaccine.

Vitespen (Oncophage), which is an HSP (gp96) vaccine, was approved by the Russian Ministry of Public Health in 2008 as an adjuvant for treatment of patients who are at intermediate risk for recurrence of RCC after nephrectomy.[41] This vaccine was approved after a phase III clinical study in patients who are at high risk of RCC recurrence. The results of 361 patients who were injected with this vaccine and 367 patients who did not receive this vaccine were analyzed in this study. The first report of this trial was published after 21 months follow-up and showed that the recurrence-free survival of both groups was statistically similar. The same results were reported after a further 17 months of follow-up and subgroup analysis.[42] The next report of this trial was published in 2009 after 3 years of follow-up. This report demonstrated that the OS of patients who received Vitespen was longer, especially among patients with intermediate risk of recurrence or earlier stages of tumor.[43]

ADVANTAGES AND DISADVANTAGES

One of the advantages of whole tumor cell vaccines is that, unlike antigen-specific vaccines, these vaccines contain characterized and uncharacterized TAAs that could be presented by APCs and stimulate appropriate immune responses.[44] Another advantage is that the MHC of these vaccines is similar to the patient's MHC. In addition, as mentioned earlier, APCs can present TAA by MHC class I and MHC class II, to the CD8+ and CD4+ T cells, respectively. Therefore, simultaneous presentation of MHC I and MHC II leads to more potent immune response against tumor.[45]

Usually a small variety of antigens that are expressed by tumor cells are tumor specific, and many of them are expressed by normal cells as well. Therefore, using whole tumor cells for vaccine preparation is not very specific and it needs to be improved.[44]

As explained earlier, whole tumor cell vaccines contain a wide variety of TAAs including unknown antigens. However, for preparation of autologous whole tumor cell vaccines, tumor biopsy is needed. However, some tumors are unreachable. Also, in some patients, the amount of obtained cells from biopsy is not sufficient or the tumor cells are necrotic.[46]

Although whole tumor cell vaccines present a wide variety of TAAs, these vaccines can release immunosuppressive cytokines[47] such as TGF-β, which leads to inhibition of developing proper immune reactions.[48]In addition, it has been shown that CD8+ T cells do not respond to tumor antigens, which are presented by MHC class I of the cancer cells due to lack of expression of costimulatory agents such as CD80 and CD86 in these cancer cells.[49]

Allogeneic whole tumor cell vaccines express TAAs. In addition, these cancer vaccines present allogeneic MHC I and MHC II, which could lead to simultaneous activation of allogeneic immune reactions and antitumor immune reactions or even rapid rejection of cancer vaccine. Simultaneous activation of immune response against allogeneic antigens with antitumor immune response can cause excessive production of cytokines. Another disadvantage of this simultaneous activation is that the activation of immune response is observed in patient's body but it is difficult to understand whether it is an allogeneic immune response or antitumor immune response.[46]

OPTIMIZATION

It has been reported that depletion of Tregs can improve the efficacy of tumor vaccines.[50,51] Tregs are a subset of CD4+ T cells with immunosuppressive activity that is controlled by the expression of forkhead box protein 3, which is a transcription factor.[52] Therefore, the efficacy of whole tumor cell vaccine can be improved by adding antiimmunosuppressive agents to these vaccines.

Another way to improve the effect of whole tumor cell vaccine is combining vaccine with anti-cytotoxic T lymphocyte-associated antigen-4 (CTLA-4). CTLA-4 is a receptor that is presented by T cells.[53] Stimulation of this receptor leads to inhibition of activation and proliferation of CD4+ and CD8+ T cells.[54] In a study, Yang et al. used anti-CTLA-4 antibody. They showed that inhibition of this receptor results in T cell activation and tumor regression in some patients with metastatic renal cell carcinoma.[55]

As mentioned earlier, CD8+ T cells do not respond to tumor antigens, which are presented by tumor cells due to lack of costimulatory agents.[49] Therefore, using costimulatory agents such as cytokines in whole tumor cell vaccine as an adjuvant can optimize these vaccines. For instance, Chiang et al. showed that combining whole tumor cell vaccine with IL-12 was more effective.[56] Another study that was conducted by Dranoff et al. reported that administration of GM-CSF to tumor cell vaccine can improve the outcomes.[57]

Also it has been shown that using radiotherapy and chemotherapy in conjunction with tumor vaccines can increase the efficacy of vaccines by using several approaches such as depletion of Tregs, elevation of TAAs, and enhancing cytokines.[58]

REFERENCES

1. Toes REM, et al. CD4 T cells and their role in antitumor immune responses. *J Exp Med* 1999;**189**(5):753–6.
2. Kelly RJ, Giaccone G. Lung cancer – vaccines. *Cancer J* 2011;**17**(5):302–8.
3. Hatfield P, et al. Optimization of dendritic cell loading with tumor cell lysates for cancer immunotherapy. *J Immunother* 2008;**31**(7):620–32.
4. Chiang CL, et al. Day-4 myeloid dendritic cells pulsed with whole tumor lysate are highly immunogenic and elicit potent anti-tumor responses. *PLoS One* 2011;**6**(12):e28732.

5. Bell CW, et al. The extracellular release of HMGB1 during apoptotic cell death. *Am J Physiol Cell Physiol* 2006;**291**(6):C1318–25.
6. Dumitriu IE, et al. Requirement of HMGB1 and RAGE for the maturation of human plasmacytoid dendritic cells. *Eur J Immunol* 2005;**35**(7):2184–90.
7. Kuppner MC, et al. The role of heat shock protein (hsp70) in dendritic cell maturation: hsp70 induces the maturation of immature dendritic cells but reduces DC differentiation from monocyte precursors. *Eur J Immunol* 2001;**31**(5):1602–9.
8. Wang Y, et al. Uric acid enhances the antitumor immunity of dendritic cell-based vaccine. *Sci Rep* 2015;**5**: 16427.
9. Andre F, et al. Malignant effusions and immunogenic tumour-derived exosomes. *Lancet* 2002;**360**(9329): 295–305.
10. Mears R, et al. Proteomic analysis of melanoma-derived exosomes by two-dimensional polyacrylamide gel electrophoresis and mass spectrometry. *Proteomics* 2004;**4**(12):4019–31.
11. Gastpar R, et al. Heat shock protein 70 surface-positive tumor exosomes stimulate migratory and cytolytic activity of natural killer cells. *Cancer Res* 2005;**65**(12):5238–47.
12. Hegmans JP, et al. Proteomic analysis of exosomes secreted by human mesothelioma cells. *Am J Pathol* 2004; **164**(5):1807–15.
13. Andreola G, et al. Induction of lymphocyte apoptosis by tumor cell secretion of FasL-bearing microvesicles. *J Exp Med* 2002;**195**(10):1303–16.
14. Lindquist S, Craig EA. The heat-shock proteins. *Annu Rev Genet* 1988;**22**:631–77.
15. Wood CG, Mulders P. Vitespen: a preclinical and clinical review. *Future Oncol* 2009;**5**(6):763–74.
16. Hoos A, Levey DL. Vaccination with heat shock protein-peptide complexes: from basic science to clinical applications. *Expert Rev Vaccines* 2003;**2**(3):369–79.
17. Rosenberg SA. A new era for cancer immunotherapy based on the genes that encode cancer antigens. *Immunity* 1999;**10**(3):281–7.
18. Teng MN, et al. Long-term inhibition of tumor growth by tumor necrosis factor in the absence of cachexia or T-cell immunity. *Proc Natl Acad Sci USA* 1991;**88**(9):3535–9.
19. Kusumoto M, et al. Phase 1 clinical trial of irradiated autologous melanoma cells adenovirally transduced with human GM-CSF gene. *Cancer Immunol Immunother* 2001;**50**(7):373–81.
20. Golumbek PT, et al. Treatment of established renal cancer by tumor cells engineered to secrete interleukin-4. *Science* 1991;**254**(5032):713–6.
21. Antonia SJ, et al. Phase I trial of a B7-1 (CD80) gene modified autologous tumor cell vaccine in combination with systemic interleukin-2 in patients with metastatic renal cell carcinoma. *J Urol* 2002;**167**(5):1995–2000.
22. Kubler H, Vieweg J. Vaccines in renal cell carcinoma. *Semin Oncol* 2006;**33**(5):614–24.
23. Nemunaitis J, et al. Phase II study of belagenpumatucel-L, a transforming growth factor beta-2 antisense gene-modified allogeneic tumor cell vaccine in non-small-cell lung cancer. *J Clin Oncol* 2006;**24**(29): 4721–30.
24. Bazhenova L, et al. An international, multicenter, randomized, double-blind phase III study of maintenance belagenpumatucel-l in non-small cell lung cancer (NSCLC): updated analysis of patients enrolled within 12 weeks of completion of chemotherapy. *J Clin Oncol* 2014;**32**(15 Suppl.). 8056–8056.
25. Lipson EJ, et al. Safety and immunologic correlates of melanoma GVAX, a GM-CSF secreting allogeneic melanoma cell vaccine administered in the adjuvant setting. *J Transl Med* 2015;**13**:214.
26. Salgia R, et al. Vaccination with irradiated autologous tumor cells engineered to secrete granulocyte-macrophage colony-stimulating factor augments antitumor immunity in some patients with metastatic non-small-cell lung carcinoma. *J Clin Oncol* 2003;**21**(4):624–30.
27. Nemunaitis J, et al. Granulocyte–macrophage colony-stimulating factor gene-modified autologous tumor vaccines in non–small-cell lung cancer. *J Natl Cancer Inst* 2004;**96**(4):326–31.

28. http://www.clinicaltrial.gov/ct2/show/NCT00089856.
29. http://www.clinicaltrial.gov/ct2/show/NCT00133224.
30. Simons JW, et al. Bioactivity of autologous irradiated renal cell carcinoma vaccines generated by ex vivo granulocyte-macrophage colony-stimulating factor gene transfer. *Cancer Res* 1997;**57**(8):1537–46.
31. Burkhardt UE, et al. Autologous CLL cell vaccination early after transplant induces leukemia-specific T cells. *J Clin Invest* 2013;**123**(9):3756–65.
32. Fakhrai H, et al. Phase I clinical trial of a TGF-beta antisense-modified tumor cell vaccine in patients with advanced glioma. *Cancer Gene Ther* 2006;**13**(12):1052–60.
33. Eton O, et al. Active immunotherapy with ultraviolet B-irradiated autologous whole melanoma cells plus DETOX in patients with metastatic melanoma. *Clin Cancer Res* 1998;**4**(3):619–27.
34. Vaishampayan U, et al. Active immunotherapy of metastatic melanoma with allogeneic melanoma lysates and interferon alpha. *Clin Cancer Res* 2002;**8**(12):3696–701.
35. Sondak VK, et al. Significant impact of HLA class I allele expression on outcome in melanoma patients treated with an allogeneic melanoma cell lysate vaccine. Final analysis of SWOG-9035. *J Clin Oncol* 2004; **22**(14 Suppl.). 7501–7501.
36. https://cms.palladianpartners.com/cms/1156354418/home.htm.
37. Dai S, et al. Phase I clinical trial of autologous ascites-derived exosomes combined with GM-CSF for colorectal cancer. *Mol Ther* 2008;**16**(4):782–90.
38. von Eschen KB, Mitchell MS. Phase III trial of melacine® melanoma theraccine versus combination chemotherapy in the treatment of stage IV melanoma: 179. *Melanoma Res* 1997;**7**:S51.
39. Mitchell MS, et al. Randomized trial of an allogeneic melanoma lysate vaccine with low-dose interferon Alfa-2b compared with high-dose interferon Alfa-2b for resected stage III cutaneous melanoma. *J Clin Oncol* 2007;**25**(15):2078–85.
40. Sosman JA, et al. Adjuvant immunotherapy of resected, intermediate-thickness, node-negative melanoma with an allogeneic tumor vaccine: impact of HLA class I antigen expression on outcome. *J Clin Oncol* 2002; **20**(8):2067–75.
41. Carlson B. Research, conferences, and FDA actions. *Biotechnol Healthc* 2008;**5**(1):7–16.
42. Wood C, et al. An adjuvant autologous therapeutic vaccine (HSPPC-96; vitespen) versus observation alone for patients at high risk of recurrence after nephrectomy for renal cell carcinoma: a multicentre, open-label, randomised phase III trial. *Lancet* 2008;**372**(9633):145–54.
43. Wood CG, et al. Survival update from a multicenter, randomized, phase III trial of vitespen versus observation as adjuvant therapy for renal cell carcinoma in patients at high risk of recurrence. *J Clin Oncol* 2009; **27**(15S). 3009–3009.
44. Lokhov PG, Balashova EE. Cellular cancer vaccines: an update on the development of vaccines generated from cell surface antigens. *J Cancer* 2010;**1**:230–41.
45. Chen M, et al. A whole-cell tumor vaccine modified to express fibroblast activation protein induces antitumor immunity against both tumor cells and cancer-associated fibroblasts. *Sci Rep* 2015;**5**:14421.
46. Yannelli JR, Wroblewski JM. On the road to a tumor cell vaccine: 20 years of cellular immunotherapy. *Vaccine* 2004;**23**(1):97–113.
47. Sheikhi A, et al. Whole tumor cell vaccine adjuvants: comparing IL-12 to IL-2 and IL-15. *Iran J Immunol* 2016;**13**(3):148–66.
48. Diaz-Valdes N, et al. Induction of monocyte chemoattractant protein-1 and interleukin-10 by TGFbeta1 in melanoma enhances tumor infiltration and immunosuppression. *Cancer Res* 2011;**71**(3):812–21.
49. Cardoso A, et al. Pre-B acute lymphoblastic leukemia cells may induce T-cell anergy to alloantigen. *Blood* 1996;**88**(1):41–8.
50. Comes A, et al. CD25+ regulatory T cell depletion augments immunotherapy of micrometastases by an IL-21-secreting cellular vaccine. *J Immunol* 2006;**176**(3):1750–8.

51. Dannull J, et al. Enhancement of vaccine-mediated. *J Clin Invest* 2005;**115**(12):3623–33.
52. Hori S, Nomura T, Sakaguchi S. Control of regulatory T cell development by the transcription factor Foxp3. *Science* 2003;**299**(5609):1057–61.
53. Chambers CA, et al. CTLA-4-mediated inhibition in regulation of T cell responses: mechanisms and manipulation in tumor immunotherapy. *Ann Rev Immunol* 2001;**19**:565–94.
54. Tivol EA, et al. Loss of CTLA-4 leads to massive lymphoproliferation and fatal multiorgan tissue destruction, revealing a critical negative regulatory role of CTLA-4. *Immunity* 1995;**3**(5):541–7.
55. Yang JC, et al. Ipilimumab (anti-CTLA4 antibody) causes regression of metastatic renal cell cancer associated with enteritis and hypophysitis. *J Immunother* 2007;**30**(8):825–30.
56. Chiang CL, et al. Optimizing parameters for clinical-scale production of high IL-12 secreting dendritic cells pulsed with oxidized whole tumor cell lysate. *J Transl Med* 2011;**9**:198.
57. Dranoff G, et al. Vaccination with irradiated tumor cells engineered to secrete murine granulocyte-macrophage colony-stimulating factor stimulates potent, specific, and long-lasting anti-tumor immunity. *Proc Natl Acad Sci USA* 1993;**90**(8):3539–43.
58. Hodge JW, et al. The tipping point for combination therapy: cancer vaccines with radiation, chemotherapy, or targeted small molecule inhibitors. *Semin Oncol* 2012;**39**(3):323–39.

CHAPTER

PEPTIDE AND PROTEIN VACCINES FOR CANCER

8

Mahsa Keshavarz-Fathi[1,2,3], Nima Rezaei[3,4,5]

School of Medicine, Tehran University of Medical Sciences, Tehran, Iran[1]; Cancer Immunology Project (CIP), Universal Scientific Education and Research Network (USERN), Tehran, Iran[2]; Research Center for Immunodeficiencies, Children's Medical Center, Tehran University of Medical Sciences, Tehran, Iran[3]; Department of Immunology, School of Medicine, Tehran University of Medical Sciences, Tehran, Iran[4]; Network of Immunity in Infection, Malignancy and Autoimmunity (NIIMA), Universal Scientific Education and Research Network (USERN), Tehran, Iran[5]

PEPTIDE-BASED VACCINE

Peptide-based vaccines are one type of vaccine developed a long time ago according to understanding of stimulation of immune system through recognizing CD4+ and CD8+ epitopes found abundant as the immunogenic components of pathogens or self-cells, which could be targeted in the context of carcinogenesis. At first, killed or attenuated pathogens were mostly used as vaccines for evoking adoptive immunity, but then, targeted therapies were developed following identification of targets such as tumor antigens.[1,2]

For selection of an antigen, as a candidate for developing a peptide vaccine, first, the tumor antigens should be identified in the tumor cells, tumor microenvironment, or peripheral blood by various laboratory tests to assess the expression of peptides and proteins, such as immunohistochemistry and reverse transcription-polymerase chain reaction (RT-PCR) for mRNA expression. For the next step, immunogenic epitopes are predicted by algorithm-based approaches using computer-aided tools. The antigens are assessed for validation on immunogenicity *in vitro* and *in vivo*. Cell cultures and animals are used for testing the desired immune responses to peptide *in vitro* and *in vivo*, and after validation, they undergo assessment in clinical trials. This time-consuming method is called reverse immunology.[3,4]

Peptide vaccines are classified into short and long peptides. Short peptides contain less than 15 amino acids and can be processed by all nucleated cells, which possess major histocompatibility complex (MHC) class I, and not necessarily by the professional antigen presenting cells (APCs). This usually results in immunologic tolerance due to lack of costimulatory molecules on the cells presenting the antigen. Utilizing potent adjuvants may lead to overcoming this weak immunogenicity. Short peptides typically can direct CD8+ T cells and not CD4+ T cell responses.[5] However, CD4+ T helper cells 1 (Th1), are necessary to induce cytotoxic T lymphocytes (CTLs) specific for attacking the tumor and providing sustained memory CTLs.[6,7] Th cells directed against the tumor antigen exert antitumor activities and produce IFN-γ and TNF-α.[8,9] Long peptides are mostly composed of both CD4+ and

Vaccines for Cancer Immunotherapy. https://doi.org/10.1016/B978-0-12-814039-0.00008-4

CD8+ epitopes and contain more than 20 amino acids. These peptides cannot be processed in unprofessional APCs because of their length. They undergo physiologic processes through both MHC I and MHC II molecules and bypass the mentioned drawback.[10,11]

To trigger a strong immune response against tumor, multiple epitopes should be presented by various MHCs to the T cells by either combining them or synthesis of a peptide consisting of more epitopes of interest.[12]

PROTEIN-BASED VACCINE

To overcome the poor immune response to the mono-epitope peptide vaccines, recombinant protein vaccines have been employed as a solution. They consist of several epitopes to expand various CD4+ and CD8+ clones. As antiviral vaccines, they have shown successful clinical results, and there is hope for exploiting them for other diseases including cancer. However, soluble proteins without appropriate adjuvants and delivery route have poor antigenicity. Therefore, designing the optimum delivery system and immunogenic adjuvants is of value to have a robust immune response.[13]

Application of a delivery system imitating actual pathogens would help with stimulating a real immunological process. It should be recognized by the APCs such as dendritic cells (DCs). Moreover, it should act as a protection for the antigenic protein against breakdown and damage, and launch the immunologic processing of the antigen to provoke robust adaptive, rather than innate, immunity and cytotoxic effects, rather than humoral responses.[14]

Using the most appropriate adjuvant for provoking a strong immune response is the other important factor that results in APC maturation. Various novel adjuvants are being combined with the vaccines such as different toll-like receptor agonists. There are other methods to make the proteins more immunogenic like using liposomes or hydrophobic polymeric particles as delivery vehicles.[12,14] There are successful examples of exploiting peptide and protein vaccines with optimized immunogenicity, which will be discussed later in the chapter.

MECHANISM OF ACTION

As the main tumor antigens inducing immune responses have peptide and protein structures, they are important molecules to be used for evoking immune responses against tumors. Therefore, peptide antigens are the targets of vaccination and are being used through various approaches to be delivered to professional APCs like DCs to prime and activate T lymphocytes as the principal players on the ground fight with cancer. The objective of vaccination is presentation of the antigens derived from the tumor by APCs to proceed with activation of long lasting adaptive immunity, especially cytotoxic responses. Antigens alone as peptide vaccines may not be efficient in induction of robust immune responses, therefore, utilization of adjuvants provides complementary signals to avoid immune tolerance.[15–17]

Proteins and peptides undergo proteolytic mechanism to be processed and loaded on MHC molecules. Generally, endogenous proteins are degraded through proteosomal proteolysis and bind to MHC-I. In contrast, exogenous proteins undergo lysosomal proteolysis and are presented by MHC-II. However, exogenous peptides, which are internalized by phagocytosis or endocytosis, bind to MHC-I through a phenomenon called cross-presentation, which is mediated by some subsets of DCs and is a crucial step for evoking CD8+ T cell response.[18] Indeed, receptors on the surface of DCs determine the subsequent events for the peptide processing. Mannose receptor (CD206)-dependent uptake of a protein leads to MHC-I cross-presentation, whereas DC-SIGN (CD209A) mediates lysosomal degradation and subsequently MHC-II binding.[19,20]

Following antigen loading to MHC, maturation of DCs and their trafficking to the secondary draining lymph nodes are necessary. In these lymphoid organs, MHC molecules, loaded with the antigen, connect to T cell receptors of antigen-specific naïve T cells. Second and third signals, which are required for T cell priming, are provided by the costimulatory molecules and cytokines.[21] The activated T cells then start to proliferate and migrate to the peripheral tissues to exert their effector activities.[22] The main objective of vaccine is eliciting optimal and efficient effector and memory T cells. Vaccine designed to deliver the tumor antigens to DCs provide the first signal needed for the priming. They can also provide the second and third signals by adding an immunogenic adjuvant. Various methods have been used to deliver the antigen.[23] Synthetic peptides and proteins derived from tumor antigens are administered as peptide and protein vaccines, which need the information about HLA molecules of patients and the binding capacity of the epitopes and MHC molecules.[24]

CLINICAL TRIALS

PEPTIDE-BASED VACCINE

There are many types of peptide vaccines examined in clinical trials, which have shown different results in various phases. We will review the development, immunogenicity, efficacy, and safety of some peptide vaccines.

Gp100

Glycoprotein 100 (gp100) is one of the glycoproteins in melanocytes, which includes a number of melanocytic differentiation antigens. The antigens are peptides composited of 8–10 amino acids that show immunogenic properties and induce CTL activity in melanoma.[25] It has been reported that a great number of CTL clones of a melanoma patient recognize HLA-A2 restricted antigens and in most cases gp100 antigen.[26] These antigens are not tumor specific, although they are of interest due to their overexpression in malignancy and inducing of immunogenicity.[25]

After validation by preclinical studies, gp100 was examined in a number of clinical trials to be evaluated in terms of safety, immunologic and clinical outcomes (Table 8.1). There are different sequences of gp100 peptides that could be used for immunization against tumor. In a study by Salgaller et al.,[27] three gp100 peptides were compared for *in vitro* sensitization and stimulation of peripheral blood mononuclear cells (PBMCs). Gp100 209 (ITDQVPFSY), gp100-280 (YLEPGPVTA), and gp100-154 (KTWGQYWQV) were used as immunogenic peptide vaccines. Modifications at anchor residues of gp100 209 and gp100-280, aiming to alter the HLA binding affinity,[28,29] led to improved immunologic response, i.e., greater number of antigen-specific CTLs.

Various immunologic indicators have been compared in pre- and postvaccination states including: proportion of CD4 and CD8 T cells; frequency of CD8+ T cells, which are reactive to the administered or natural peptide; their ability to produce cytotoxic substances such as IFN-γ, perforin and granzyme B, memory and effector phenotypes of T cells; frequencies of regulatory T cells, monocytes, and immature myeloid-derived cells; expression of costimulatory molecules such as CD40, CD80, and CD86; secretion of interleukins, chemokine ligands; and delayed-type hypersensitivity (DTH).[30–32]

Table 8.1 Clinical Trials Investigating gp100 Peptide Vaccine for Cancer

Study	Phase	Patients	Number (Analyzed/Included Patients)	Intervention	Control	Immunologic Outcome	Clinical Outcome	Safety
M. L. Salgaller et al.[27]	I	HLA-A*020l+ melanoma patients	28	The G9-154, G9-209, or G9-280 peptides	–	IFNγ secretion, cytotoxicity	Regression	–
S. R. Reynolds et al.[39]	I	HLA-A*02+ patients with resected malignant melanoma	22	Polyvalent vaccine containing multiple Ags, including MAGE-3, Melan-A/MART-1, gp100, tyrosinase, melanocortin receptor (MC1R), and dopachrome tautomerase (TRP-2)	–	CD8+ T cell responses, Relative responses to individual Ags	Survival, progression-free survival	
T. M. Clay et al.[34]	I	HLA-A2+ melanoma patients	3	g9-209 2M	–	T cell reactivity	–	–
C. L. Slingluff Jr. et al.[40]	I	Patients with resected stage 2B, 3, and 4 melanoma or patients with minimal metastatic disease with HLA-A2+ and tumor gp100+	22	$gp100_{280-288}$ with or without the HLA-DR-restricted tetanus helper peptide	–	T cell response, CD8+ T cell responses, IFN-γ secretion, Th1-type helper T cell responses, DTH responses	Tumor progression, survival and disease-free survival	Reported
J. W. Smith et al.[41]	I	HLA-A2+ patients with cutaneous malignant melanoma and no distant metastasis	30	gp100 209–2M with or without IFN-α	–	CD8+ T cell responses, IFN-γ secretion, effector/memory T cell response.	Recurrence	Reported
E. B. Walker et al.[42]	I	HLA-A2+ patients with completely resected stage 1–3 melanoma	35 (29 from the study by J.W. Smith et al., 2003)	gp100 209–2M peptide and a control "reporter" peptide, human papillomavirus (HPV) 16E712–20 emulsified in the Montanide ISA 51.	–	CD8+ T cell responses, CFC (IFN-γ)+CD8+ frequency, effector/memory T cell response	–	–

D. S. Cassarino et al.[43]	I	HLA A2+ patients with at least stage 2 melanoma and surgical excision of both their primary and metastatic lesions	10	gp100: 209–219 (210M) and tyrosinase: 368–396 (370D)	–	Melanocytic markers, T cell markers, CD4/8 ratio, MHC class 1 and 2 markers and Langerhans cell marker CD1a	–	–
T. Di Pucchio et al.[30]	I	Stage 4 pretreated metastatic melanoma patients with HLA A*0201+	10	Melan-A/MART-1 and gp100 peptides plus IFN-α	–	Frequency of T cells producing IFN-γ, CD8+ T cells response, phenotype of peptide-specific T cells, CD14+ monocytes, expression of CD40, CD80, and CD86 costimulatory molecules, Allostimulatory activity of patients' monocytes, IL-5, IL-10, and IP-10/CXCL10	Clinical response and disease-free interval	Side effects, signs of autoimmunity
S. L. Meijer et al.[44]	I	Stage 1–3 HLA-A2 positive patients with primary melanomas >1-mm thick undergone surgery	29	gp100 209-2M and HPV16E7: 12–20, in Montanide ISA 51 adjuvant	–	IFN-γ secretion, circulating CD8+ T, Melanoma cell lysis, Cytotoxicity assays, Clinical correlation	–	–
J. A. Sosman et al.[31]	II	Patients with advanced HLA-A2 positive melanoma	131	210M and Montanide ISA-51 plus high-dose IL-2	–	Frequency of CD8+ T cells, immature myeloid-derived cells, and Tregs and their correlation with clinical responses	Clinical responses progression-free survival, overall survival	Performance status, toxicity of high-dose IL-2

Continued

Table 8.1 Clinical Trials Investigating gp100 Peptide Vaccine for Cancer—cont'd

Study	Phase	Patients	Number (Analyzed/ Included Patients)	Intervention	Control	Immunologic Outcome	Clinical Outcome	Safety
E. B. Walker et al.[32]	I	HLA-A2+ patients with resected stage 1 to 3 melanoma	35	gp100 209-2M peptide in Montanide ISA adjuvant	–	Frequencies of peptide-specific CD8+ T cells in sentinel lymph nodes and peripheral blood, CD107 cell surface expression, phenotype of peptide-specific T cells in lymph nodes than peripheral blood	–	–
D.J. Schwartzentruber et al.[37]	III	HLA*A0201-positive patients with metastatic cutaneous melanoma, either stage IV or locally advanced stage III	185	gp100:209–217 (210M)plus incomplete Freund's adjuvant (Montanide ISA-51) and interleukin-2	Interleukin-2 alone	Anti-peptide reactivity, relationship between the development of antipeptide reactivity and the objective clinical response. Posttreatment levels of CD4+foxp3+ cells	Clinical response, progression free survival, overall survival.	Toxic effects
A. A. Tarhini et al.[45]	I	HLA-A2 + melanoma patients with stage IV	22	Multi-epitope peptide vaccine containing MART-1 (26–35, 27L), gp100 (209–217, 210M) and tyrosinase (368–376, 370D) peptides, given combined with the immunomodulators GM-CSF and PF3512676 in Montanide ISA oil adjuvant	–	Peptide reactive CD8+ T cells,	Response rate, progression free survival, overall survival.	Reported

Induction of CD8+ T cells reactive to peptide vaccine, which are CTLs is considered as an important immunologic outcome after vaccine administration and is assessed in the majority of clinical trials. Both frequency and function of CTLs account for cytotoxic effects. Owing to this fact, investigators perform ELISPOT, HLA/peptide tetramer staining, and IFN-γ release assays. These assays usually were not different before and after vaccination without *in vitro* stimulation, therefore, the native or modified peptides are used for sensitizing the PBMCs *in vitro* to induce CTLs reactive to gp100.[33] CD8+ T cell response to gp100 peptide vaccine is the most reported immunologic outcome in phase 1 clinical trials. A greater percentage of CD8+ T cells derived from PBMCs were reactive to modified gp100 209 at the second position (gp100 209/2M) compared to their reactivity to melanoma cell lines or native gp100.[34] However, further studies showed that despite the higher affinity of altered peptides and increased frequency of T cells specific for them, T cells primed by the natural peptide exhibit an enhanced avidity and effector activities.[28,35]

Perforin and granzyme B are the other cytotoxic molecules indicating the activity of CTLs. CD107 expression on the surface of T cells showed correlation with secretion of these molecules, and in one study, it was reported higher in PBMCs after stimulation in comparison with sentinel lymph node–derived T cells. Functional avidity of CD8+ T cells in peripheral blood and more fully differentiated effector T cells were also associated with higher expression of CD107.[32] However, correlation of immunologic responses with the clinical responses is required for decision-making about effectiveness of vaccines.[36]

There are a lot of trials examining gp100 for melanoma treatment, a number of which are listed in Table 8.1. In a phase 3 randomized clinical trial, subcutaneous gp100:209–217(210M) peptide and Montanide ISA-51 (once per cycle) following by intravenous interleukin-2 (IL-2) (720,000 IU per kilogram of body weight per dose) were compared with IL-2 alone at the same dose. One hundred and eighty five HLA-A0201 positive patients with stage 4 or locally advanced stage 3 melanoma were randomized to the arms of study to be evaluated for clinical response, safety, and progression-free survival (PFS). IL-2 was injected up to a maximum of 12 doses per cycle, and the cycle of therapy was repeated every 3 weeks. More adverse effects were reported in the patients who received high-dose IL-2 in the combination arm. The adverse effects were also analyzed in the first two cycles of treatment, in which the dose of IL-2 was the same between two arms, and only arrhythmias were reported with higher incidence in the combination arm. Clinical responses and PFS were improved in the combination group (clinical response, 16% vs. 6%, $P = 0.03$). There was an improvement of overall survival (OS) in the combination arm but not statistically significant (17.8 vs. 11.1 $P = 0.06$).[37]

Despite the mentioned study, a phase 3 study compared efficacy of gp100 peptide vaccine and ipilimumab, a CTLA4 inhibitor. Up to four cycle of treatment was administered every 3 weeks. Ipilimumab (3 mg per kilogram of body weight) increased OS to 10.1 compared with 6.4 months in gp100 peptide vaccine (hazard ratio for death, 0.68; $P < 0.001$). Notably, no differences were reported in OS between the ipilimumab alone versus ipilimumab combined with gp100 groups (OS 10.1 vs. 10.0 months, hazard ratio with ipilumumab combined with gp100, 1.04; $P = 0.76$).[38]

BLP25

Biomira lipopeptide 25 (BLP25) is a synthetic peptide that could trigger immunity against cancer cells expressing mucin 1 (MUC1). MUC1 is overexpressed and aberrantly glycosylated in a majority of adenocarcinomas including lung, prostate, breast, ovarian, and multiple myeloma and hepatic metastases from colorectal cancer.[46–51] It also shows expression in the normal epithelial tissues, however, its glycosylation pattern, which is altered in malignant tissue, protects the normal form from the immune system.[52] BLP25 liposomal (L- BLP25) vaccine's safety and efficacy were assessed for several tumor types and the trials are still underway. Table 8.2 summarizes the information of the trials testing L- BLP25.

L-BLP25 takes advantage of liposome as a vector. The adjuvant monophosphoryl lipid A, and three lipids, are its other components.[53] In advanced stages of lung cancer, a phase 2b trial was conducted to assess the efficacy and safety of L-BLP25 plus best supportive care (BSC) in comparison with BSC alone. In the vaccine arm, a single dose of cyclophosphamide (300 mg/m2) was administered intravenously and eight doses of L-BLP25 (1,000 microg) was administered subcutaneously each week. Maintenance doses of L-BLP25 were administered each 6-weeks. The treatment was continued until progression or withdrawal. Longer OS was reported in the combination group (4.4 months improvement, $P = 0.112$).[54] In the updated report of this trial, 17.3 months improvement of OS was yielded in a subgroup of patients with stage 3B loco-regional cancer, who received vaccine, compared with those received the BSC alone (30.6 vs. 13.3 months; HR 0.548, 95% CI 0.301–0.999).[55] In a phase 3 trial, L-BLP25 was compared with placebo in 1513 patients with stage 3 non-small-cell lung cancer (NSCLC) not progressed after the primary therapy (START trial). The treatment cycles were similar to the previous trial of L-BLP25 but at 806 microg. There was a trend in favor of the vaccine in improvement of OS (25.6 vs. 22.3 months; adjusted HR 0.88, 95% CI: 0.75–1.03, $P = 0.123$). A subgroup of patients who received concurrent chemotherapy and radiation had an improved OS on the L-BLP25 than the placebo (30.8 vs. 20.6 months; HR 0.78, 95% CI: 0.64–0.95, $P = 0.016$).[56] Hence, based on these results, modification was applied to the two ongoing trials on L-BLP25, i.e., START2 and INSPIRE.[57,58]

PROTEIN BASED VACCINE

MAGE-A3

Melanoma-associated antigen 3 (MAGE-A3) protein is a member of cancer-testis antigens (CTAs), which is expressed in 35%–50% of NSCLC, 74% of melanoma, 35% of sarcoma and bladder carcinoma specimens, and lower in some other types of cancer.[66–68] CTAs are antigens expressed in embryogenesis period, in testis, during germ cell development, and in placenta.[69] However, they are also overexpressed in several types of tumor and could be presented by both HLA class I and class II, which make them an antigen of interest to trigger both arms of adaptive immunity. Interestingly, HLA class I and class II are not expressed on testes or placenta.[70,71] Hence, the vaccine does not trigger immune responses against these organs. MAGE-A3 as a tumor-specific antigen is a candidate to be targeted for active specific cancer immunotherapy.

In a double-blind randomized phase 2 clinical trial, 182 patients with MAGE-A3 positive NSCLC post resection, received this protein plus GlaxoSmithKline's immunostimulant AS02 B ($n = 122$) or placebo ($n = 60$). Totally, 13 doses of vaccine were administered intramuscularly including first five doses every 3 weeks and next eight doses every 3 months. After median follow up of 44 months, 43 (35%) patients in treatment group and 26 (43%) in the placebo group showed tumor relapse. Hazard ratios for disease-free interval (DFI), disease-free survival (DFS), and OS were in favor of the treatment group, but not statistically significant. The treatment was well tolerated with 14% and 13% grade 3 or 4 toxicity in the treatment and control group, respectively, which were apparently not different.

Table 8.2 Clinical Trials Investigating BLP25 Liposomal Vaccine for Cancer

Study	Phase	Patients	Number (Analyzed/ Included Patients)	Intervention	Control	Immunologic Outcome	Clinical Outcome	Safety
M. Palmer et al.[59]	1	Stage 3B or stage 4 non-small-cell lung cancer (NSCLC)	17	BLP25 liposomal vaccine	–	CTL activity, antibody assay, lymphoproliferation	Clinical response, progression free survival, overall survival	Reported
S. A. North et al.[60]	1	Prostate cancer patients with biochemical failure after radical prostatectomy	16	BLP25 liposome vaccine	–	–	PSA level, PSADT	Reported
C. Butts et al.[54]	2	Patients with stage 3B and 4 NSCLC	171	BLP25 liposome vaccine plus best supportive care (BSC)	Best supportive care (BSC)	MUC1-specific T cell proliferative response	Overall survival, quality of life	Adverse events related to cyclophosphamide and BLP25
C. Butts et al.[61]	1	Patients with unresectable stage 3 non–small-cell lung cancer	22	BLP25 liposome vaccine, cyclophosphamide, and BSC	–	–	1-year survival rate, the 2-year survival rate	Adverse events related to cyclophosphamide and BLP25
F. Ohyanagi et al.[62]	1	Japanese patients with unresectable stage 3 NSCLC after primary chemoradiotherapy	6	BLP25 liposome vaccine	–	Serum concentrations of IL-1b, sIL-2 Ra, IL-6, IL-8, and TNFa	–	Reported
C. Butts et al.[55]	2	Updated data of the C. Butts et al., 2005 study	171	BLP25 liposome vaccine plus BSC (BSC)	BSC (BSC)	–	Overall survival, 3-year survival rate	–
Y. L. Wu et al.[58]	3	Asian patients with unresectable stage 3 NSCLC	420	BLP25 liposome vaccine	Placebo	Not reported yet	Not reported yet	Not completed yet

Continued

Table 8.2 Clinical Trials Investigating BLP25 Liposomal Vaccine for Cancer—cont'd

Study	Phase	Patients	Number (Analyzed/ Included Patients)	Intervention	Control	Immunologic Outcome	Clinical Outcome	Safety
C. C. Schimanski et al.[51]	2	Patients with stage 4 colorectal carcinoma after curative resection of primary tumor and hepatic metastases	159	BLP25 liposome vaccine	Placebo	Not reported yet	Not reported yet	Not completed yet
C. Butts et al.[63]	3	Patients with stage 3 unresectable NSCLC	1513	BLP25 liposome vaccine	Placebo	–	Overall survival, 1-year survival rate, 2-year survival rate, 3-year survival rate, Time to symptom progression, median time to disease progression	Reported
E. Rossmann et al.[64]	1	Patients with asymptomatic stage 1/2 or stage 2/3 multiple myeloma in stable response/plateau phase	34	BLP25 liposome vaccine with either single or repeated cyclophosphamide	–	MUC1-specific immune response, frequencies of different phenotypes of T cell, cytokine production	Time to tumor progression, progressive disease rate, M-protein concentration	Adverse events related to cyclophosphamide and BLP25
P. Mitchell et al.[65]	3	Updated data of the C. Butts et al., 2014 study	1513	BLP25 liposome vaccine	Placebo	sMUC1, ANA, lymphocyte count, neutrophil/ lymphocyte ratio, HLA typing	Overall survival	–

All patients developed antibodies against MAGE-A3 mostly after four doses of vaccination. However, there was no association between antibody response and DFI.[72]

The most renowned clinical trial investigating protein vaccines was MAGRIT, which evaluated MAGE-A3 as an adjuvant for antigen-specific cancer immunotherapy. It was aimed to assess tumor relapse prevention in patients with proved MAGE-A3 positive NSCLC, who underwent surgery.[73] In the MAGRIT study, 1515 patients received MAGE-A3 plus AS15 immunostimulant and 757 patients received placebo. The treatment regimen was similar to the previous phase II trial. After median follow-up of 38.1–39.5 months, median DFS was 60.5 months for MAGE-A3 and 57.9 months for the placebo group, but not statistically significant (HR 1.02, 95% CI 0.89–1.18; $P = 0.74$).[74]

EGF-P64k

EGF-P64K or CIMAvax-EGF is another protein vaccine, which targets the epidermal growth factor (EGF). EGF is an extracellular protein ligand, which binds its receptor, i.e., EGFR, and leads to proliferation and survival of tumor cells.[75] More than half of NSCLC specimens (40%–80%) over-express EGFR.[76] To block the signaling pathways activated following the interaction of EGF and EGFR, EGF-P64 K or CIMAvax-EGF is used to generate anti-EGF antibodies and impede their interaction.[77] This vaccine contains the protein P64k, originate from *Neisseria meningitides*, as a carrier to deliver EGF. Montanide ISA 51 is used as the adjuvant as well. The vaccine has been evaluated in a number of phase 1 and 2 trials and demonstrated promising results and led to conducting a phase 3 trial.[78]

In the phase 3 trial, 405 patients with advanced NSCLC were randomized to receive either CIMAvax-EGF or the control, BSC. Safety, immunogenicity, and EGF levels in sera of patients as well as OS were assessed. The vaccine was safe and induced humoral response against EGF, which led to decreased level of this protein in the serum. In the vaccine arm, a single dose of cyclophosphamide (200 mg/m2) was administered intravenously and four doses of vaccine was administered intramuscularly every 2 weeks and then every month. The OS improved in the vaccine arm but was not statistically significant when using the standard log rank (10.83 vs. 8.86 months; HR, 0.82; $P = 0.100$). In case of the confirmation of the nonproportionality of the HR, a weighted log-rank test can be applied. In this trial, Harrington–Fleming was applied and resulted in the significant differences between the OS of two arms ($P = 0.04$). In the patients, who completed the vaccination with at least four doses, a greater survival advantage was observed in the vaccine arm than the BSC arm (12.43 vs. 9.43 months; HR, 0.77; $P = 0.036$).[79]

ADVANTAGES AND DISADVANTAGES

Vaccines are a type of biologic drug showing fewer side effects than chemical substances and radiation. Among them, peptide- and protein-based vaccines overall might be the safest, unless the toxicity that is induced by their adjuvants used to evoke more robust cell-based immunity. Hence, they usually are safe and well-tolerated and do not lead to autoimmunity. They are small molecules and, after identification, fewer steps must be undertaken for their production and manufacturing compared to complex technologies used for majority of other vaccines. There are standard methods for their production and preparation, which make them proper candidates for global production and usage. The other benefit is low cost of manufacturing that makes them affordable and more available all around the world for both developed and developing countries. Taken together, there are enough benefits for peptide- and protein-based treatments including vaccines, and they are good candidates for development of off-the-shelf drugs.[80,81]

To address some of the demerits of this type of vaccine, the identification and selection of appropriate tumor peptides are technically challenging. Since longer peptides are more immunogenic and their structure alters their immunogenicity, selection of the proper peptide and sometimes making modifications are substantial for designing an efficient vaccine. Computer-aided tools used for prediction of epitopes might show false-negative and false-positive results as well.[82,83]

Peptides are recognized in an MHC-restricted manner, which could affect their off-the-shelf usage. If they are mono-epitope vaccines, usually they do not contain CD4+ epitopes and cannot trigger Th cells. However, selecting a mixture of peptides presented by the MHC molecules, more prevalent in humans, and utilization of different epitopes bypasses this limit. Short peptides might also induce tolerance and be degraded in the body. The immunogenicity and efficacy of peptide vaccines in clinical trials are another subject of importance, which could be improved by some strategies and combination therapy.[80,84,85]

OPTIMIZATION

Optimization is undertaken to improve immunologic and clinical efficacy of a therapeutic approach. To optimize peptide vaccines, elimination of obstacles in a number of areas, which influence efficacy, such as selection of appropriate peptide, combination of immunogenic epitopes, using polypeptide vaccines, choice of correct adjuvant, delivery system, and route of administration, are undeniable. Moreover, the dose of peptide vaccine and the boosting schedule play an important role in the outcomes of using peptide vaccines. High frequencies of antigen-specific T cells and their functional avidity should be kept balanced. Higher dose of peptide leading to increased antigen presentation by DCs is a method to increase the number of antigen-specific T cells. However, the peptide should be administered in an adjusted dose in the case of further vaccination or in case of first vaccination while the peptide is processed by nonprofessional APCs. In these cases, higher doses of peptides lead to activation of T cells with lower avidity and vice versa. The structure of peptide can also affect the result of vaccination. Sometimes alteration in the structure leads to altered T cell responses, which are not appropriately predictable.[86]

REFERENCES

1. Reche P, et al. Peptide-based immunotherapeutics and vaccines 2015. *J Immunol Res* 2015;**2015**:349049.
2. Boohaker RJ, et al. The use of therapeutic peptides to target and to kill cancer cells. *Curr Med Chem* 2012; **19**(22):3794–804.
3. Hickman ES, Lomax ME, Jakobsen BK. Antigen selection for enhanced affinity T-cell receptor–based cancer therapies. *J Biomol Screen* 2016;**21**(8):769–85.
4. Rane SS, Javad JMS, Rees RC. Tumor antigen and epitope identification for preclinical and clinical evaluation. In: Rezaei N, editor. *Cancer immunology: bench to bedside immunotherapy of cancers*. Berlin, Heidelberg: Springer Berlin Heidelberg; 2015. p. 55–71.
5. Slingluff CL. The present and future of peptide vaccines for cancer: single or multiple, long or short, alone or in combination? *Cancer J (Sudbury, Mass)* 2011;**17**(5):343–50.

6. Janssen EM, et al. CD4+ T cells are required for secondary expansion and memory in CD8+ T lymphocytes. *Nature* 2003;**421**(6925):852–6.
7. Bos R, Sherman LA. CD4+ T-cell help in the tumor milieu is required for recruitment and cytolytic function of CD8+ T lymphocytes. *Cancer Res* 2010;**70**(21):8368–77.
8. Braumuller H, et al. T-helper-1-cell cytokines drive cancer into senescence. *Nature* 2013;**494**(7437):361–5.
9. Xie Y, et al. Naive tumor-specific CD4(+) T cells differentiated in vivo eradicate established melanoma. *J Exp Med* 2010;**207**(3):651–67.
10. Zom GG, et al. Efficient induction of antitumor immunity by synthetic toll-like receptor ligand-peptide conjugates. *Cancer Immunol Res* 2014;**2**(8):756–64.
11. Melief CJ, van der Burg SH. Immunotherapy of established (pre)malignant disease by synthetic long peptide vaccines. *Nat Rev Cancer* 2008;**8**(5):351–60.
12. Yang H, Kim DS. Peptide immunotherapy in vaccine development: from epitope to adjuvant. *Adv Protein Chem Struct Biol* 2015;**99**:1–14.
13. Nascimento IP, Leite LCC. Recombinant vaccines and the development of new vaccine strategies. *Braz J Med Biol Res* 2012;**45**(12):1102–11.
14. Beitelshees M, Li Y, Pfeifer BA. Enhancing vaccine effectiveness with delivery technology. *Curr Opin Biotechnol* 2016;**42**:24–9.
15. Ying H, Zeng G, Black KL. Innovative cancer vaccine strategies based on the identification of tumour-associated antigens. *BioDrugs* 2001;**15**(12):819–31.
16. Palucka K, Banchereau J. Dendritic-cell-based therapeutic cancer vaccines. *Immunity* 2013;**39**(1):38–48.
17. Azmi F, et al. Recent progress in adjuvant discovery for peptide-based subunit vaccines. *Hum Vaccines Immunother* 2014;**10**(3):778–96.
18. Blum JS, Wearsch PA, Cresswell P. Pathways of antigen processing. *Annu Rev Immunol* 2013;**31**: 443–73.
19. Li Y, et al. Tumor-derived autophagosome vaccine: mechanism of cross-presentation and therapeutic efficacy. *Clin Cancer Res* 2011;**17**(22):7047–57.
20. Cheong C, et al. Microbial stimulation fully differentiates monocytes to DC-SIGN/CD209(+) dendritic cells for immune T cell areas. *Cell* 2010;**143**(3):416–29.
21. Khong H, Overwijk WW. Adjuvants for peptide-based cancer vaccines. *J Immunother Cancer* 2016;**4**(1):56.
22. Krummel MF, Bartumeus F, Gérard A. T-cell migration, search strategies and mechanisms. *Nat Rev Immunol* 2016;**16**(3):193–201.
23. Andrews DM, Maraskovsky E, Smyth MJ. Cancer vaccines for established cancer: how to make them better? *Immunol Rev* 2008;**222**:242–55.
24. González FE, et al. Tumor cell lysates as immunogenic sources for cancer vaccine design. *Hum Vaccines Immunother* 2014;**10**(11):3261–9.
25. Vigneron N, et al. A peptide derived from melanocytic protein gp100 and presented by HLA-B35 is recognized by autologous cytolytic T lymphocytes on melanoma cells. *Tissue Antigens* 2005;**65**(2): 156–62.
26. Zarour H, et al. The majority of autologous cytolytic T-lymphocyte clones derived from peripheral blood lymphocytes of a melanoma patient recognize an antigenic peptide derived from gene Pmel17/gp100. *J Invest Dermatol* 1996;**107**(1):63–7.
27. Salgaller ML, et al. Immunization against epitopes in the human melanoma antigen gp100 following patient immunization with synthetic peptides. *Cancer Res* 1996;**56**(20):4749–57.
28. Cole DK, et al. Modification of MHC anchor residues generates heteroclitic peptides that alter TCR binding and T cell recognition. *J Immunol* 2010;**185**(4):2600–10.
29. Bianchi V, et al. A molecular switch abrogates glycoprotein 100 (gp100) T-cell receptor (TCR) targeting of a human melanoma antigen. *J Biol Chem* 2016;**291**(17):8951–9.

30. Di Pucchio T, et al. Immunization of stage IV melanoma patients with Melan-A/MART-1 and gp100 peptides plus IFN-alpha results in the activation of specific CD8(+) T cells and monocyte/dendritic cell precursors. *Cancer Res* 2006;**66**(9):4943–51.
31. Sosman JA, et al. Three phase II cytokine working group trials of gp100 (210M) peptide plus high-dose interleukin-2 in patients with HLA-A2-positive advanced melanoma. *J Clin Oncol* 2008;**26**(14):2292–8.
32. Walker EB, et al. Characterization of the class I-restricted gp100 melanoma peptide-stimulated primary immune response in tumor-free vaccine-draining lymph nodes and peripheral blood. *Clin Cancer Res* 2009; **15**(7):2541–51.
33. Parkhurst MR, et al. Improved induction of melanoma-reactive CTL with peptides from the melanoma antigen gp100 modified at HLA-A*0201-binding residues. *J Immunol* 1996;**157**(6):2539–48.
34. Clay TM, et al. Changes in the fine specificity of gp100(209-217)-reactive T cells in patients following vaccination with a peptide modified at an HLA-A2.1 anchor residue. *J Immunol* 1999;**162**(3):1749–55.
35. Speiser DE, et al. Unmodified self antigen triggers human CD8 T cells with stronger tumor reactivity than altered antigen. *Proc Natl Acad Sci USA* 2008;**105**(10):3849–54.
36. Butterfield LH. Cancer vaccines. *BMJ* 2015;**350**:h988.
37. Schwartzentruber DJ, et al. gp100 peptide vaccine and interleukin-2 in patients with advanced melanoma. *N Engl J Med* 2011;**364**(22):2119–27.
38. Hodi FS, et al. Improved survival with ipilimumab in patients with metastatic melanoma. *N Engl J Med* 2010; **363**(8):711–23.
39. Reynolds SR, et al. HLA-independent heterogeneity of CD8+ T cell responses to MAGE-3, Melan-A/MART-1, gp100, tyrosinase, MC1R, and TRP-2 in vaccine-treated melanoma patients. *J Immunol* 1998; **161**(12):6970–6.
40. Slingluff Jr CL, et al. Phase I trial of a melanoma vaccine with gp100(280-288) peptide and tetanus helper peptide in adjuvant: immunologic and clinical outcomes. *Clin Cancer Res* 2001;**7**(10):3012–24.
41. Smith 2nd JW, et al. Adjuvant immunization of HLA-A2-positive melanoma patients with a modified gp100 peptide induces peptide-specific CD8+ T-cell responses. *J Clin Oncol* 2003;**21**(8):1562–73.
42. Walker EB, et al. gp100(209-2M) peptide immunization of human lymphocyte antigen-A2+ stage I-III melanoma patients induces significant increase in antigen-specific effector and long-term memory CD8+ T cells. *Clin Cancer Res* 2004;**10**(2):668–80.
43. Cassarino DS, et al. The effects of gp100 and tyrosinase peptide vaccinations on nevi in melanoma patients. *J Cutan Pathol* 2006;**33**(5):335–42.
44. Meijer SL, et al. Induction of circulating tumor-reactive CD8+ T cells after vaccination of melanoma patients with the gp100 209-2M peptide. *J Immunother* 2007;**30**(5):533–43.
45. Tarhini AA, et al. Safety and immunogenicity of vaccination with MART-1 (26-35, 27L), gp100 (209-217, 210M), and tyrosinase (368-376, 370D) in adjuvant with PF-3512676 and GM-CSF in metastatic melanoma. *J Immunother* 2012;**35**(4):359–66.
46. Wong N, et al. Amplification of MUC1 in prostate cancer metastasis and CRPC development. *Oncotarget* 2016;**7**(50):83115–33.
47. Patel DS, et al. Immunohistochemical study of MUC1, MUC2 and MUC5AC expression in primary breast carcinoma. *J Clin Diagn Res* 2017;**11**(4):Ec30–ec34.
48. Situ D, et al. Expression and prognostic relevance of MUC1 in stage IB non-small cell lung cancer. *Med Oncol* 2011;**28**(Suppl. 1):S596–604.
49. Van Elssen CH, et al. Expression of aberrantly glycosylated Mucin-1 in ovarian cancer. *Histopathology* 2010; **57**(4):597–606.
50. Takahashi T, et al. Expression of MUC1 on myeloma cells and induction of HLA-unrestricted CTL against MUC1 from a multiple myeloma patient. *J Immunol* 1994;**153**(5):2102–9.

51. Schimanski CC, et al. LICC: L-BLP25 in patients with colorectal carcinoma after curative resection of hepatic metastases—a randomized, placebo-controlled, multicenter, multinational, double-blinded phase II trial. *BMC Cancer* 2012;**12**. 144—144.
52. Brayman M, Thathiah A, Carson DD. MUC1: a multifunctional cell surface component of reproductive tissue epithelia. *Reprod Biol Endocrinol* 2004;**2**:4.
53. Sangha R, Butts C. L-BLP25: a peptide vaccine strategy in non small cell lung cancer. *Clin Cancer Res* 2007; **13**(15 Pt 2). s4652-4.
54. Butts C, et al. Randomized phase IIB trial of BLP25 liposome vaccine in stage IIIB and IV non-small-cell lung cancer. *J Clin Oncol* 2005;**23**(27):6674—81.
55. Butts C, et al. Updated survival analysis in patients with stage IIIB or IV non-small-cell lung cancer receiving BLP25 liposome vaccine (L-BLP25): phase IIB randomized, multicenter, open-label trial. *J Cancer Res Clin Oncol* 2011;**137**(9):1337—42.
56. Butts CA, et al. START: a phase III study of L-BLP25 cancer immunotherapy for unresectable stage III non-small cell lung cancer. *J Clin Oncol* 2013;**31**(15 Suppl.). 7500—7500.
57. Ramlogan-Steel CA, Steel JC, Morris JC. Lung cancer vaccines: current status and future prospects. *Transl Lung Cancer Res* 2014;**3**(1):46—52.
58. Wu YL, et al. INSPIRE: a phase III study of the BLP25 liposome vaccine (L-BLP25) in Asian patients with unresectable stage III non-small cell lung cancer. *BMC Cancer* 2011;**11**:430.
59. Palmer M, et al. Phase I study of the BLP25 (MUC1 peptide) liposomal vaccine for active specific immunotherapy in stage IIIB/IV non—small-cell lung cancer. *Clin Lung Cancer* 2001;**3**(1):49—57.
60. North S, Butts C. Vaccination with BLP25 liposome vaccine to treat non-small cell lung and prostate cancers. *Expert Rev Vaccines* 2005;**4**(3):249—57.
61. Butts C, et al. A multicenter open-label study to assess the safety of a new formulation of BLP25 liposome vaccine in patients with unresectable stage III non-small-cell lung cancer. *Clin Lung Cancer* 2010;**11**(6): 391—5.
62. Ohyanagi F, et al. Safety of BLP25 liposome vaccine (L-BLP25) in Japanese patients with unresectable stage III NSCLC after primary chemoradiotherapy: preliminary results from a Phase I/II study. *Jpn J Clin Oncol* 2011;**41**(5):718—22.
63. Butts C, et al. Tecemotide (L-BLP25) versus placebo after chemoradiotherapy for stage III non-small-cell lung cancer (START): a randomised, double-blind, phase 3 trial. *Lancet Oncol* 2014;**15**(1):59—68.
64. Rossmann E, et al. Mucin 1-specific active cancer immunotherapy with tecemotide (L-BLP25) in patients with multiple myeloma: an exploratory study. *Hum Vaccine Immunother* 2014;**10**(11):3394—408.
65. Mitchell P, et al. Tecemotide in unresectable stage III non-small-cell lung cancer in the phase III START study: updated overall survival and biomarker analyses. *Ann Oncol* 2015;**26**(6):1134—42.
66. Sienel W, et al. Melanoma associated antigen (MAGE)-A3 expression in Stages I and II non-small cell lung cancer: results of a multi-center study. *Eur J Cardio Thorac Surg* 2004;**25**(1):131—4.
67. Brasseur F, et al. Expression of MAGE genes in primary and metastatic cutaneous melanoma. *Int J Cancer* 1995;**63**(3):375—80.
68. Peled N, et al. MAGE A3 antigen-specific cancer immunotherapeutic. *Immunotherapy* 2009;**1**(1):19—25.
69. Jungbluth AA, et al. Expression of cancer-testis (CT) antigens in placenta. *Cancer Immunol* 2007;**7**:15.
70. Kurpisz M, et al. Hla expression on human germinal cells. *Int J Immunogenet* 1987;**14**(1):23—32.
71. Blaschitz A, Hutter H, Dohr G. HLA Class I protein expression in the human placenta. *Early Pregnancy* 2001;**5**(1):67—9.
72. Vansteenkiste J, et al. Adjuvant MAGE-A3 immunotherapy in resected non-small-cell lung cancer: phase II randomized study results. *J Clin Oncol* 2013;**31**(19):2396—403.
73. Tyagi P, Mirakhur B. MAGRIT: the largest-ever phase III lung cancer trial aims to establish a novel tumor-specific approach to therapy. *Clin Lung Cancer* 2009;**10**(5):371—4.

74. Vansteenkiste JF, et al. Efficacy of the MAGE-A3 cancer immunotherapeutic as adjuvant therapy in patients with resected MAGE-A3-positive non-small-cell lung cancer (MAGRIT): a randomised, double-blind, placebo-controlled, phase 3 trial. *Lancet Oncol* 2016;**17**(6):822–35.
75. Avraham R, Yarden Y. Feedback regulation of EGFR signalling: decision making by early and delayed loops. *Nat Rev Mol Cell Biol* 2011;**12**:104.
76. Toyooka S, et al. Molecular oncology of lung cancer. *Gen Thorac Cardiovasc Surg* 2011;**59**(8):527–37.
77. Garcia B, et al. Effective inhibition of the epidermal growth factor/epidermal growth factor receptor binding by anti-epidermal growth factor antibodies is related to better survival in advanced non-small-cell lung cancer patients treated with the epidermal growth factor cancer vaccine. *Clin Cancer Res* 2008;**14**(3):840–6.
78. Rodriguez PC, et al. Clinical development and perspectives of CIMAvax EGF, Cuban vaccine for non-small-cell lung cancer therapy. *MEDICC Rev* 2010;**12**(1):17–23.
79. Rodriguez PC, et al. A phase III clinical trial of the epidermal growth factor vaccine CIMAvax-EGF as switch maintenance therapy in advanced non-small cell lung cancer patients. *Clin Cancer Res* 2016;**22**(15): 3782–90.
80. Kumai T, et al. Peptide vaccines in cancer-old concept revisited. *Curr Opin Immunol* 2017;**45**:1–7.
81. Hirayama M, Nishimura Y. The present status and future prospects of peptide-based cancer vaccines. *Int Immunol* 2016;**28**(7):319–28.
82. Fridman A, Finnefrock AC, Peruzzi D, et al. An efficient T-cell epitope discovery strategy using in silico prediction and the iTopia assay platform. *Oncoimmunology* 2012;**1**(8):1258–70. https://doi.org/10.4161/onci.21355.
83. Karasaki T, Nagayama K, Kuwano H, et al. Prediction and prioritization of neoantigens: integration of RNA sequencing data with whole-exome sequencing. *Cancer Sci* 2017;**108**(2):170–7. https://doi.org/10.1111/cas.13131.
84. Li W, et al. Peptide vaccine: progress and challenges. *Vaccines* 2014;**2**(3):515–36.
85. Guo C, et al. Therapeutic cancer vaccines: past, present and future. *Adv Cancer Res* 2013;**119**:421–75.
86. Leggatt GR. Peptide dose and/or structure in vaccines as a determinant of T cell responses. *Vaccines* 2014; **2**(3):537–48.

CHAPTER

IMMUNE CELL VACCINE FOR CANCER

Sepideh Razi[1,2], Nima Rezaei[3,4,5]

Cancer Immunology Project (CIP), Universal Scientific Education and Research Network (USERN), Tehran, Iran[1]; Student Research Committee, School of Medicine, Iran University of Medical Sciences, Tehran, Iran[2]; Research Center for Immunodeficiencies, Children's Medical Center, Tehran University of Medical Sciences, Tehran, Iran[3]; Department of Immunology, School of Medicine, Tehran University of Medical Sciences, Tehran, Iran[4]; Network of Immunity in Infection, Malignancy and Autoimmunity (NIIMA), Universal Scientific Education and Research Network (USERN), Tehran, Iran[5]

IMMUNE CELL VACCINE

Allogeneic or autologous immune cells have been widely used for development of cancer vaccines. Among different types of immune cells, dendritic cells (DCs) have shown the most success in cancer vaccination. Therefore, this chapter will focus on DC-based cancer vaccines.

MECHANISM OF ACTION

Adaptive and innate immunity both play important roles in responses against cancer. Innate immune cells such as natural killer (NK) cells could directly detect and eliminate tumor cells.[1] Whereas adaptive immunity is antigen specific, antigen presenting cells (APCs) acquire tumor antigens from circulating proteins and dying tumor cells. Tumor antigens are processed to shorter peptides, which are linked to the major histocompatibility complex (MHC) molecules and are presented on extracellular surface of APCs. Peptide fragments, which are in conjunction with MHC class I molecules, are recognized by CD8+ cytotoxic T lymphocytes (CTL). CTLs can cause cell death by releasing lytic enzymes or Fas/Fas ligand pathway, which leads to apoptosis.[2] CD4+ T cells are activated by MHC class II complexes on APCs. CD4+ T cells could facilitate the activation of CD8+ T cells. Activation of CD4+ T cells leads to elevation of the cytotoxic activity of NK cells. In addition, CD4+ T cells could provoke hormonal immunity and antibody production.[3]

DC is one of the professional APCs that have a significant role in relation between adaptive and innate immune responses[4] through the activation of different cells of immune system such as CD4+, CD8+, and B cells. DCs are naturally derived from CD14+ monocytes and bone marrow CD34+ cells.[5,6] These cells are present in peripheral tissues and can travel to lymphoid organs to prime T cell

Vaccines for Cancer Immunotherapy. https://doi.org/10.1016/B978-0-12-814039-0.00009-6

responses by presenting different antigens.[7] It has been shown that one of the escape mechanisms of tumor cells from immunity is inadequate presentation of tumor-associated antigens (TAAs) to naïve lymphocytes.[8] In addition, it has been shown that infiltration of DCs into primary tumor sites is associated with longer survivals in a wide variety of cancers.[9] Therefore, broad varieties of cancer vaccines have been developed by using these cells. For development of these vaccines, leukapheresis is used to collect bone marrow CD34+ cells or CD14+ monocytes from patient's blood (autologous) or another person (allogeneic) for preparation of DCs and then these cells are cultured with growth macrophage colony stimulating factor (GM-CSF) and other kinds of proinflammatory cytokines to produce immature DCs. However, because of low efficacy of DC-based vaccines, which consist of immature DCs, some methods have been utilized to provide mature DCs. Factors such as CD40 ligand, lipopolysaccharide, tumor necrosis factor alpha (TNFα), interferon alpha (IFN-α), and gamma (IFN-γ) have been used for maturation of immature DCs.[10,11] These DC-based vaccines can be used either alone or in combination with cytokine-induced killer cells. However, one of the disadvantages of these vaccines is that these vaccines do not activate tumor-specific immune responses. In addition, these vaccines are short-lived. Therefore, they do not activate memory cells.[12] Therefore, in order to stimulate immune responses against tumor specific antigens, DCs are pulsed with TAAs such as tumor cell lysates, tumor cell peptides, and tumor-derived RNA.

CLINICAL TRIALS

IMMATURE DENDRITIC CELLS PULSED WITH TUMOR CELL LYSATES

A phase I clinical study was carried out to test the effect of autologous DCs pulsed with autologous tumor cell lysates, which contain a broad variety of TAAs, in patients with metastatic renal cell carcinoma (mRCC). In this trial, monocytes were cultured with IL-4 and GM-CSF to prepare immature autologous DCs. These immature DCs were loaded with autologous tumor cell lysates and then were given to a total of 12 patients. In 6 out of 12 of these participants, DCs were also loaded with keyhole limpet hemocyanin (KLH). In addition, all of the patients were given low-dose interleukin-2 (IL-2). In this study, KLH reactivity was observed. However, none of these patients showed antibody production and other immune reactions against tumor. However, extended stable disease (SD) was observed. Therefore, the results showed that immature DCs pulsed with autologous tumor lysates had no significant effect in patients with mRCC.[13] In Another study by Azuma et al., three patients with RCC received immature DCs loaded with autologous tumor cell lysates and KLH. Two out of three patients showed T cell immune reaction, SD was observed in one patient, and all of the participants showed KLH reactivity. Local injection reaction was observed in all patients but no serious toxicities were observed.[14] In addition, another study was performed to test the efficacy of a DC-based vaccine that consisted of immature DCs pulsed with tumor cell lysates in mRCC patients. The results of 12 patients were analyzed and showed that one patient had a partial response (PR) and three patients showed SD. No significant change in immune responses against tumor was reported. This vaccine was well tolerated and side effects were restricted to grade 1.[15] In another study, five patients with RCC received immature DCs loaded with allogeneic tumor cell lysates and KLH. Two out of five patients showed SD and one patient showed delayed hypersensitivity reactions (DTHs) to KLH. In addition, no significant side effect was observed in this study.[16]

MATURE DENDRITIC CELLS PULSED WITH TUMOR CELL LYSATES

In a phase I/II clinical study, a total of 15 patients with mRCC were given autologous DCs loaded with tumor cell lysates. Also, DCs were pulsed with KLH in three patients. The proliferation rate of peripheral blood lymphocytes and the number of CD3+CD28+ and CD3+CD4+ cells noticeably upregulated in the patients' blood. PR was observed in one patient, seven patients showed progressive disease (PD), and seven patients showed SD. Significant adverse effects were not observed in this trial.[17] Another clinical study used DCs pulsed with tumor cell lysates in 12 patients with mRCC. The results showed the activation of immune responses against tumor lysates and KLH. In addition, some lesions in the lungs of one patient were remitted and only moderate fever was reported as a side effect.[18] Also, this vaccine was tested in another study of 35 patients with mRCC. The results of 27 patients were assessed. Two out of 27 patients had complete response, 7 patients showed SD, 1 patient had a PR, and the other 11 patients showed PD. The activation of immune responses was evaluated in 11 patients. All 11 patients showed increased immune reactions. Flu-like symptoms were the only side effect that was observed in this study.[19]

The median overall survival (OS) of patients with glioblastoma is approximately 15 months. Therefore, it is very important to use a treatment that can improve the survival of these patients. A phase II clinical study was conducted to test the effect of using autologous DCs pulsed with whole tumor cell lysates in patients with glioblastoma in combination with resection of the tumor, radiotherapy, and temozolomide, which is a chemotherapeutic drug. For vaccine development, autologous DCs were provided by using peripheral blood monocytes of the patients and then pulsed with whole tumor cell lysates. After resection of the tumor, patients with residual tumor volume of less than 1 cc were enrolled in this study. Vaccination of the patients started before radiotherapy and was pursued during treatment with temozolomide. The results of the study showed that the median OS of the patients was 23.4 months and the median progression-free survival (PFS) was 12.7 months. Moreover, antitumor-specific immune reactions after vaccination were observed in 11 out of 27 patients who were assessed. However, the results did not show any correlation between increased antitumor immune reactions and patients' survival. No major adverse effects were reported.[20] A phase II clinical study tested the effect of autologous DCs loaded with tumor cell lysates in 44 patients (34 patients were diagnosed with glioblastoma). Fifty-three percent of glioblastoma patients showed $\geq$ 1.5-fold vaccine-enhanced cytokine responses. In addition, vaccine responders showed longer survival compared with vaccine nonresponders. No grades 3 or 4 toxicity were observed in this trial.[21] Another phase I/II study was conducted in 77 patients with glioblastoma. The median OS was 18.3 months and no serious side effects were observed in this study. This study suggested that this vaccine is safe and that combination of this vaccine with surgery, chemotherapy, and radiotherapy can be useful for treatment of glioblastoma.[22] Additionally, at present, there is an ongoing clinical study investigating the effect of DC vaccines pulsed with tumor cell lysates for the treatment of malignant glioma and glioblastoma (NCT01808820).[23]

A clinical trial was carried out to investigate the effect of a DC-based vaccine in patients with stage IV melanoma. DCs were produced by using the peripheral blood monocytes of the patients and then pulsed with heterologous tumor cell lysates. A total of 11 patients with melanoma were enrolled in this trial and received this vaccine monthly in an inguinal lymph node that was normal according to ultrasound evaluation. The results showed a PR for 5 months in one patient. In addition, a mixed response was reported in two patients. The median OS was 7.3 months. The activation of immune responses against tumor was observed and no major side effects were reported.[24]

The efficacy of a DC-based vaccine was tested in a phase I/II clinical study in patients with advanced soft tissue and bone sarcoma. Peripheral blood mononuclear cells were collected from patients and then cultured with GM-CSF and IL-4. After 7 days, DCs were treated with autologous tumor cell lysates, OK-432 and TNFα. Patients received six doses of this vaccine weekly in axillary or inguinal region. The results of 37 patients were assessed in this study and showed that the serum level of IL-12 and IFN-γ were increased 1 month after vaccination. In addition, the clinical responses of 35 patients were analyzed. Eight weeks after vaccination, 28 out of 35 patients had tumor progression, 6 patients showed SD, and 1 patient had a PR. The 3-year OS rate of the patients was 42.3% and the PFS rate was 2.9%. This vaccine was well tolerated and no major adverse effects were observed. Therefore, according to the results, only a small population of patients showed an improvement in clinical responses.[25]

DENDRITIC CELLS LOADED WITH TUMOR-DERIVED RNA

A phase I clinical trial was carried out to test the effect of DCs loaded with renal tumor RNA in 10 patients with mRCC. Activation of T cell immune response against a wide variety of TAAs such as G250, telomerase reverse transcriptase (hTERT), and oncofetal antigen were reported. Evaluation of clinical responses in this study was not possible due to receiving other treatment. However, only 3 out of 10 patients died because of tumor after 19.8 months of follow-up and cancer-related mortality was low. No major toxicity was reported in this study.[26] In another clinical study of 10 patients with RCC, DCs were loaded with RNA of autologous tumor tissue and each patient received three vaccinations. Also six patients received IL-2 diphtheria toxin conjugate before vaccination in order to decrease the number of CD4+CD25+ regulatory T cells (Tregs). The results reported that patients who received IL-2 diphtheria toxin conjugate before vaccination showed noticeably upregulated CD8+ T cell response. In addition, the results demonstrated elevation of CD8+ T cell reactions and IFN-γ-secreting CD4+ T cells in nine patients. Clinical responses were not shown in this study.[27]

The efficacy of a DC-based vaccine, which was loaded with messenger RNA (mRNA), was investigated in patients with melanoma. In a clinical trial, 30 melanoma patients received autologous mRNA-electroporated DCs after resection of metastasis. However, patients had micrometastasis at the time of vaccination. The 2-year survival rate was 93% and 4-year survival rate was 70%. The side effects of this vaccine were restricted to local injection site reactions in all participants, skin depigmentation in seven cases and one patient with grade 2 flu-like symptoms and fever.[28] Another phase I/II clinical study was conducted to test the efficacy of a vaccine consisting of mature autologous DCs pulsed with mRNA of tumor. A total of 22 patients with malignant melanoma were enrolled in this study and received these vaccines weekly for 1 month. The activation of immune responses against tumor was observed in 7 out of 10 patients who received these vaccines intradermally and 3 out of 12 patients who were given intranodal injection of these vaccines. The stimulation of immune responses against tumor was evaluated in 9 out of 19 patients by using T cell assays. In addition, DTH was reported in 8 out of 18 patients who were assessed. The study did not report any significant side effects. Therefore, according to the results of the trial, this vaccine can trigger the activation of antitumor T cell responses *in vivo*, and patients who received intradermal injection of this vaccine showed better results.[29]

A clinical study investigated the effect of DCs pulsed with tumor RNA and KLH in 15 patients with colorectal cancer. The results showed a decrease in carcinoembryonic antigen (CEA) of 7 patients, and the skin test for KLH was positive in 11 out of 13 patients who were evaluated. No serious adverse effects were reported.[30]

DENDRITIC CELLS PULSED WITH PEPTIDES

A clinical study was performed to investigate the effect of DCs pulsed with MUC-1 peptides vaccine in 20 patients with mRCC. Patients received this vaccine every 2 weeks for 8 weeks and then, until tumor progression, this schedule was repeated. In addition, after the fifth injection, patients were given weekly injection of low-dose IL-2 for 3 weeks. The T cell immune responses against MUC-1 peptide were observed in 6 patients. In addition, remission of metastatic sites in six patients were reported. SD was observed in 4 patients up to 14 months. No major side effects were reported in this trial.[31] Another phase I study was conducted in patients with mRCC to investigate the effect of DCs pulsed with carbonic anhydrase-IX (CA9) peptide and KLH. Patients were given five vaccinations. All of the participants showed anti-KLH humeral responses and DTH. There was no evidence of the activation of immune responses against CA-9 peptide. Also, no clinical responses were reported. Adverse effects of this vaccine were not significant.[32]

The efficacy of a DC-based vaccine that was loaded with peptides was evaluated in a phase I clinical trial in 4/7 patients with glioma. Two out of 4 patients were reoperated after vaccination. These 2 patients showed robust infiltration of CD8+ and memory T cells. Systemic cytotoxicity was observed in 7 patients.[33] In another study in 12 patients with glioblastoma that were given DCs pulsed with acid-eluted peptides, 6 out of 12 patients showed systemic CTL reactions against tumor and 4 out of 8 patients who were reoperated, showed elevated numbers of tumor infiltrating lymphocytes. No major adverse effects were observed in this trial.[34]

The effect of DC-based vaccines, which were pulsed with peptides, was investigated in patients with melanoma. For instance, a clinical study was performed in patients with melanoma. Patients received DCs pulsed with a different variety of peptides, which can be recognized by CD8+ CTLs, or with tumor cell lysates. Eleven out of 16 patients showed DTH. In addition, CTLs responses against peptides were observed. Five out of 16 patients showed regression of metastatic lesions in multiple organs.[35] In a phase I clinical study, a total of 16 patients with melanoma were given DCs loaded with gp100 and tyrosinase peptides intravenously. One patient showed complete remission (CR), 2 out of 16 patients showed SD, and 2 patients showed mix responses.[36] In another clinical study, DCs pulsed with gp100, tyrosinase, MAGE-3 and MelanA/MART-1 melanoma peptides, as well as KLH and influenza matrix as control antigens were injected subcutaneously to 18 patients with melanoma. Sixteen out of 18 patients showed immune reactions against control antigens and 10 patients showed immune responses to at least two peptides of melanoma.[37]

A DC-based vaccine was tested in a phase I/II clinical study in patients with colorectal cancer. Mature DCs were generated by using interferon-α, OK432, and prostanoid and pulsed with carcinoembryonic antigen (CEA) peptide. Patients received four doses of this vaccine every 2 weeks. The results of 8 patients were analyzed. One patient showed SD and the other 7 patients had PD. In addition, the results showed elevated frequency and cytotoxic activity of NK cells in the patients who had SD. Increased activation of CTLs against CEA antigen was also observed in this patient.

In patients who showed PD, the proliferation of NK cells was reported. This vaccine was well tolerated and only grade 1 fever and induration at injection sites were reported in all patients.[38]

DENDRITIC CELLS LOADED WITH TUMOR CELLS

A phase II clinical study in patients with metastatic melanoma investigated the effect of a DC-based vaccine. In order to prepare this vaccine, DCs were pulsed with autologous tumor cells. DTH was reported in 1 out of 54 patients at baseline versus 12 out of 54 patients after receiving vaccination. Additionally, following 4.5 years of follow-up, the 5-year survival rate was 54%. This trial did not report any major toxicity and the side effects were restricted to local injection site reactions such as edema and pruritus.[39] Another phase II clinical study was conducted in the patients with melanoma. Patients were randomly divided into two groups. One group received irradiated autologous proliferating tumor cells and the patients in the other group were given DCs pulsed with antigens of autologous tumor cells. The results showed that the 2-year survival rates were 72% for patients who received DC-based vaccine versus 31% for the patients who received tumor cell vaccine ($P = 0.007$). Therefore, this trial suggested the use of DC-based vaccine loaded with tumor antigens could lead to a longer survival compared with tumor cell vaccine.[40]

A clinical study investigated the effect of a DC-based vaccine in patients with non-small cell lung cancer (NSCLC). DCs were produced by using CD14+ precursors and then were loaded with apoptotic bodies from an allogeneic NSCLC cell line that overexpressed CEA, Wilms tumor 1 (WT1), survivin, Her2/neu, and MAGE-2. A total of 16 patients with stage 1A to 3B NSCLC were enrolled in this trial. Patients were given two doses of vaccine intradermally. The results showed the activation of specific immune responses against tumor in 6 out of 16 patients. Also, 5 out of 16 patients showed the activation of immune responses but these responses were not dependent to tumor antigens. In addition, there was no evidence of the activation of immune responses in 5 out of 16 patients. This vaccine was well tolerated and no major side effects were observed.[41]

APPROVED VACCINE

Sipuleucel-T (Provenge) is the first and currently the only US Food and Drug Administration–approved cell-based cancer vaccine. In order to prepare this vaccine, DCs are collected by leukapheresis from the patient's blood and then cultured with a recombinant fusion protein, which consists of prostatic acid phosphatase (PAP) and GM-CSF. Patients' T cells are activated against PAP when they receive this vaccine. Therefore, T cells can recognize and induce death in PAP overexpressed prostate cancer cells. This vaccine was approved after a phase III clinical trial in 2010. In this trial, a total of 512 patients with metastatic castrate-resistant prostate cancer were enrolled. The results showed that the median OS of patients who received this vaccine was 25.8 versus 21.7 months for placebo. However, similar clinical progression in both groups was reported.[42]

ADVANTAGES AND DISADVANTAGES

As mentioned earlier, one of the escape mechanisms of tumors from immune responses is insufficient presentation of TAAs to naïve lymphocytes.[8] Therefore, one of the advantages of DC-based vaccines is presenting enough TAAs to immune system. Another advantage is that there is still a large variety of TAAs that are unknown. Therefore, by using DCs pulsed with tumor cell lysates or tumor cells, patients' immune responses are activated against different kinds of TAAs, some of which are still uncharacterized.[43] In addition, it has been shown that DC-based vaccines are safer than other immunotherapy methods. In addition, use of these vaccines can lead to long-term antitumor memory.[44]

One of the disadvantages of the DC-based vaccines loaded with tumor TAAs is that it can trigger autoimmune responses against normal cell antigens by presenting housekeeping or lineage-associated epitopes.[45] In addition, in the process of developing DC-based vaccines that are loaded with tumor cells or TAAs, tumor specimens are needed, but in some patients tumors are not reachable or in some cases obtained tumor biopsy is insufficient. Another problem in the process of preparation of these vaccines is the failure of *in vitro* tumor cell culture. A further disadvantage of DC-based vaccines is that these vaccines are patient specific and all of the processes in the preparation of these vaccines are time and cost consuming.[44]

As mentioned earlier, a disadvantage of using DC-based vaccines, which consist of immature DCs, is that these immature DCs can cause T cell anergy instead of T cell activation. Also they can induce the differentiation of regulatory T cells.[46]

OPTIMIZATION

Targeting the immunosuppressive agents such as IL-10,[47] 1L-6,[48] Tregs,[49] and myeloid-derived suppressor cells (MDSCs)[50] in tumor microenvironment can increase the efficacy of DC-based vaccines. For instance, it has been shown that Treg suppression can lead to better efficacy. A clinical study tested the effect of DC-based vaccine in combination with 1-methyl-D-tryptophan (1-D-MT), which can suppress Tregs in patients with breast cancer (NCT01042535).[51] In addition, Tregs express IL-2 receptors, and currently a clinical study is investigating the effect of basiliximab, a monoclonal antibody that can target these receptors, in patients with glioblastoma (NCT02366728).[52] MDSCs are another example of immunosuppressive agents that can be inhibited by using several substances such as COX-2 inhibitors and vitamin D3.[53]

Another strategy to improve the efficacy of DC-based vaccines is checkpoint inhibitors administration. It has been shown that when PD-1, which is a checkpoint protein on T cells, binds to PD-ligand 1 (PD-L1), which is presented by different cells such as DCs, it can lead to PD-1 activation, which results in T cell anergy.[54] Zhang et al. used PD-L1 signaling inhibitors in combination with DC-based vaccines in breast tumor—bearing hu-SCID mice. The results showed more favorable outcomes than using DC-based vaccines alone.[55]

Another way to optimize DC-based vaccines is preparing adjuvants, which can increase the numbers and/or activity of DCs. For instance, toll-like receptors (TLRs) are membrane receptors that can be stimulated by binding to pathogen-associated molecular patterns (PAMPs), and it has been shown that TLR signaling can trigger DC maturation.[56] Therefore, TLR agonists can be used as adjuvants to DC-based vaccines.

Radiotherapy of tumor cells can lead to cell death. Necrotic cells are a source of TAAs. In addition, radiation therapy of tumor cells results in elevated expression of MHC class I. Therefore, increased presentation of TAAs by necrotic cells and elevated presentation of MHC class I can increase the immunogenicity of tumor cells.[57,58] Radiotherapy induces tumor cells to release high-mobility group protein B1 (HMGB1), which binds to TLR-4 and activates DCs.[59] Therefore, using radiotherapy in combination with DC-based vaccines can improve the efficacy of the vaccines.

Different chemotherapeutic drugs can optimize the effect of these vaccines by several mechanisms, including inducing tumor cells to upregulate immunogenicity, activating T cells, and stimulating immune reactions to decrease tumor-induced immunosuppressive agents and cells.[60]

It has been shown that the delivery route of DC-based vaccines can affect the efficacy of these vaccines. There are several routes of administration of vaccines such as subcutaneous, intratumoral, intradermal, intranodal, and intravenous.[61] In a study, Lesterhuis et al. compared the efficacy of a DC-based vaccine between the patients who received this vaccine intranodally versus patients who were given the intradermal administration of this vaccine. The activation of T cell responses against tumor were observed in 19 out of 43 patients. Elevated migration of DCs to the lymph nodes was observed in patients who received intranodal vaccination. However, no evidence of migration of DCs to the lymph nodes was reported in 7 out of 24 patients who were given intranodal vaccination. In addition, this study reported stronger activation of immune responses against tumor in patients who received intradermal vaccination.[62] In another study, peptide-pulsed DC vaccines were given to patients with metastatic melanoma. Patients were randomly divided into four groups. Patients in the first group received intranodal injection of the vaccine. Patients in the second and third groups were given intravenous and intradermal vaccination. The results showed increased activation of CD8+ T cells in all patients. However, the activation of these cells was higher in patients who received intranodal injection.[63]

REFERENCES

1. Marcus A, et al. Recognition of tumors by the innate immune system and natural killer cells. *Adv Immunol* 2014;**122**:91–128.
2. Kelly RJ, Giaccone G. Lung cancer vaccines. *Cancer J* 2011;**17**(5):302–8.
3. Toes REM, et al. CD4 T cells and their role in antitumor immune responses. *J Exp Med* 1999;**189**(5):753–6.
4. Steinman RM, Cohn ZA. Identification of a novel cell type in peripheral lymphoid organs of mice. I. Morphology, quantitation, tissue distribution. *J Exp Med* 1973;**137**(5):1142–62.
5. Bernhard H, et al. Generation of immunostimulatory dendritic cells from human CD34+ hematopoietic progenitor cells of the bone marrow and peripheral blood. *Cancer Res* 1995;**55**(5):1099–104.
6. Romani N, et al. Generation of mature dendritic cells from human blood. An improved method with special regard to clinical applicability. *J Immunol Methods* 1996;**196**(2):137–51.
7. Kirk CJ, Hartigan-O'Connor D, Mule JJ. The dynamics of the T-cell antitumor response: chemokine-secreting dendritic cells can prime tumor-reactive T cells extranodally. *Cancer Res* 2001;**61**(24):8794–802.
8. Gilboa E. How tumors escape immune destruction and what we can do about it. *Cancer Immunol Immunother* 1999;**48**(7):382–5.
9. Lotze MT. Getting to the source: dendritic cells as therapeutic reagents for the treatment of patients with cancer. *Ann Surg* 1997;**226**(1):1–5.
10. Murthy V, et al. Clinical considerations in developing dendritic cell vaccine based immunotherapy protocols in cancer. *Curr Mol Med* 2009;**9**(6):725–31.
11. Castiello L, et al. Monocyte-derived DC maturation strategies and related pathways: a transcriptional view. *Cancer Immunol Immunother* 2011;**60**(4):457–66.
12. Tesfatsion DA. Dendritic cell vaccine against leukemia: advances and perspectives. *Immunotherapy* 2014;**6**(4):485–96.

13. Oosterwijk-Wakka JC, et al. Vaccination of patients with metastatic renal cell carcinoma with autologous dendritic cells pulsed with autologous tumor antigens in combination with interleukin-2: a phase 1 study. *J Immunother* 2002;**25**(6):500–8.
14. Azuma T, et al. Dendritic cell immunotherapy for patients with metastatic renal cell carcinoma: University of Tokyo experience. *Int J Urol* 2002;**9**(6):340–6.
15. Gitlitz BJ, et al. A pilot trial of tumor lysate-loaded dendritic cells for the treatment of metastatic renal cell carcinoma. *J Immunother* 2003;**26**(5):412–9.
16. Pandha HS, et al. Dendritic cell immunotherapy for urological cancers using cryopreserved allogeneic tumour lysate-pulsed cells: a phase I/II study. *BJU Int* 2004;**94**(3):412–8.
17. Marten A, et al. Therapeutic vaccination against metastatic renal cell carcinoma by autologous dendritic cells: preclinical results and outcome of a first clinical phase I/II trial. *Cancer Immunol Immunother* 2002; **51**(11–12):637–44.
18. Holtl L, et al. Cellular and humoral immune responses in patients with metastatic renal cell carcinoma after vaccination with antigen pulsed dendritic cells. *J Urol* 1999;**161**(3):777–82.
19. Holtl L, et al. Immunotherapy of metastatic renal cell carcinoma with tumor lysate-pulsed autologous dendritic cells. *Clin Cancer Res* 2002;**8**(11):3369–76.
20. Inoges S, et al. A phase II trial of autologous dendritic cell vaccination and radiochemotherapy following fluorescence-guided surgery in newly diagnosed glioblastoma patients. *J Transl Med* 2017;**15**(1):104.
21. Wheeler CJ, et al. Vaccination elicits correlated immune and clinical responses in glioblastoma multiforme patients. *Cancer Res* 2008;**68**(14):5955–64.
22. Ardon H, et al. Integration of autologous dendritic cell-based immunotherapy in the standard of care treatment for patients with newly diagnosed glioblastoma: results of the HGG-2006 phase I/II trial. *Cancer Immunol Immunother* 2012;**61**(11):2033–44.
23. https://clinicaltrials.gov/ct2/show/NCT01808820?term=NCT01808820&rank=1.
24. Vilella R, et al. Pilot study of treatment of biochemotherapy-refractory stage IV melanoma patients with autologous dendritic cells pulsed with a heterologous melanoma cell line lysate. *Cancer Immunol Immunother* 2004;**53**(7):651–8.
25. Miwa S, et al. Phase 1/2 study of immunotherapy with dendritic cells pulsed with autologous tumor lysate in patients with refractory bone and soft tissue sarcoma. *Cancer* 2017;**123**(9):1576–84.
26. Su Z, et al. Immunological and clinical responses in metastatic renal cancer patients vaccinated with tumor RNA-transfected dendritic cells. *Cancer Res* 2003;**63**(9):2127–33.
27. Dannull J, et al. Enhancement of vaccine-mediated. *J Clin Invest* 2005;**115**(12):3623–33.
28. Wilgenhof S, et al. Long-term clinical outcome of melanoma patients treated with messenger RNA-electroporated dendritic cell therapy following complete resection of metastases. *Cancer Immunol Immunother* 2015;**64**(3):381–8.
29. Kyte JA, et al. Phase I/II trial of melanoma therapy with dendritic cells transfected with autologous tumor-mRNA. *Cancer Gene Ther* 2006;**13**(10):905–18.
30. Rains N, et al. Development of a dendritic cell (DC)-based vaccine for patients with advanced colorectal cancer. *Hepato-Gastroenterology* 2001;**48**(38):347–51.
31. Wierecky J, et al. Immunologic and clinical responses after vaccinations with peptide-pulsed dendritic cells in metastatic renal cancer patients. *Cancer Res* 2006;**66**(11):5910–8.
32. Bleumer I, et al. Preliminary analysis of patients with progressive renal cell carcinoma vaccinated with CA9-peptide-pulsed mature dendritic cells. *J Immunother* 2007;**30**(1):116–22.
33. Yu JS, et al. Vaccination of malignant glioma patients with peptide-pulsed dendritic cells elicits systemic cytotoxicity and intracranial T-cell infiltration. *Cancer Res* 2001;**61**(3):842–7.

34. Liau LM, et al. Dendritic cell vaccination in glioblastoma patients induces systemic and intracranial T-cell responses modulated by the local central nervous system tumor microenvironment. *Clin Cancer Res* 2005; **11**(15):5515–25.
35. Nestle FO, et al. Vaccination of melanoma patients with peptide- or tumor lysate-pulsed dendritic cells. *Nat Med* 1998;**4**(3):328–32.
36. Lau R, et al. Phase I trial of intravenous peptide-pulsed dendritic cells in patients with metastatic melanoma. *J Immunother* 2001;**24**(1):66–78.
37. Banchereau J, et al. Immune and clinical responses in patients with metastatic melanoma to CD34(+) progenitor-derived dendritic cell vaccine. *Cancer Res* 2001;**61**(17):6451–8.
38. Sakakibara M, et al. Comprehensive immunological analyses of colorectal cancer patients in the phase I/II study of quickly matured dendritic cell vaccine pulsed with carcinoembryonic antigen peptide. *Cancer Immunol Immunother* 2011;**60**(11):1565–75.
39. Dillman RO, et al. Phase II trial of dendritic cells loaded with antigens from self-renewing, proliferating autologous tumor cells as patient-specific antitumor vaccines in patients with metastatic melanoma: final report. *Cancer Biother Radiopharm* 2009;**24**(3):311–9.
40. Dillman RO, et al. Tumor stem cell antigens as consolidative active specific immunotherapy: a randomized phase II trial of dendritic cells versus tumor cells in patients with metastatic melanoma. *J Immunother* 2012; **35**(8):641–9.
41. Hirschowitz EA, et al. Autologous dendritic cell vaccines for non-small-cell lung cancer. *J Clin Oncol* 2004; **22**(14):2808–15.
42. Kantoff PW, et al. Sipuleucel-T immunotherapy for castration-resistant prostate cancer. *N Engl J Med* 2010; **363**(5):411–22.
43. Yoshimura K, Uemura H. Role of vaccine therapy for renal cell carcinoma in the era of targeted therapy. *Int J Urol* 2013;**20**(8):744–55.
44. Javed A, Sato S, Sato T. Autologous melanoma cell vaccine using monocyte-derived dendritic cells (NBS20/eltrapuldencel-T). *Future Oncol* 2016;**12**(6):751–62.
45. Morisaki T, et al. Dendritic cell-based combined immunotherapy with autologous tumor-pulsed dendritic cell vaccine and activated T cells for cancer patients: rationale, current progress, and perspectives. *Hum Cell* 2003;**16**(4):175–82.
46. Granucci F, et al. Dendritic cell regulation of immune responses: a new role for interleukin 2 at the intersection of innate and adaptive immunity. *EMBO J* 2003;**22**(11):2546–51.
47. de Vries JE. Immunosuppressive and anti-inflammatory properties of interleukin 10. *Ann Med* 1995;**27**(5): 537–41.
48. Tilg H, Dinarello CA, Mier JW. IL-6 and APPs: anti-inflammatory and immunosuppressive mediators. *Immunol Today* 1997;**18**(9):428–32.
49. Sakaguchi S, et al. Regulatory T cells: how do they suppress immune responses? *Int Immunol* 2009;**21**(10): 1105–11.
50. Gabrilovich DI, Nagaraj S. Myeloid-derived-suppressor cells as regulators of the immune system. *Nat Rev Immunol* 2009;**9**(3):162–74.
51. https://clinicaltrials.gov/ct2/show/results/NCT01042535?term=NCT01042535&rank=1.
52. https://clinicaltrials.gov/ct2/results?cond=&term=NCT02366728&cntry=&state=&city=&dist=.
53. Najjar YG, Finke JH. Clinical perspectives on targeting of myeloid derived suppressor cells in the treatment of cancer. *Front Oncol* 2013:3.
54. Dong Y, Sun Q, Zhang X. PD-1 and its ligands are important immune checkpoints in cancer. *Oncotarget* 2017;**8**(2):2171–86.
55. Ge Y, et al. Blockade of PD-1/PD-L1 immune checkpoint during DC vaccination induces potent protective immunity against breast cancer in hu-SCID mice. *Cancer Lett* 2013;**336**(2):253–9.

56. López CB, Yount JS, Moran TM. Toll-like receptor-independent triggering of dendritic cell maturation by viruses. *J Virol* 2006;**80**(7):3128–34.
57. Kwilas AR, et al. In the field: exploiting the untapped potential of immunogenic modulation by radiation in combination with immunotherapy for the treatment of cancer. *Front Oncol* 2012:2.
58. Hauser SH, et al. Radiation-enhanced expression of major histocompatibility complex class I antigen H-2Db in B16 melanoma cells. *Cancer Res* 1993;**53**(8):1952–5.
59. Apetoh L, et al. Toll-like receptor 4-dependent contribution of the immune system to anticancer chemotherapy and radiotherapy. *Nat Med* 2007;**13**(9):1050–9.
60. Weir GM, Liwski RS, Mansour M. Immune modulation by chemotherapy or immunotherapy to enhance cancer vaccines. *Cancers* 2011;**3**(3):3114–42.
61. Black M, et al. Advances in the design and delivery of peptide subunit vaccines with a focus on toll-like receptor agonists. *Expert Rev Vaccines* 2010;**9**(2):157–73.
62. Lesterhuis WJ, et al. Route of administration modulates the induction of dendritic cell vaccine-induced antigen-specific T cells in advanced melanoma patients. *Clin Cancer Res* 2011;**17**(17):5725–35.
63. Bedrosian I, et al. Intranodal administration of peptide-pulsed mature dendritic cell vaccines results in superior CD8+ T-cell function in melanoma patients. *J Clin Oncol* 2003;**21**(20):3826–35.

CHAPTER

GENETIC VACCINE FOR CANCER 10

Saeed Farajzadeh Valilou[1], Mahsa Keshavarz-Fathi[1,2,3]

Cancer Immunology Project (CIP), Universal Scientific Education and Research Network (USERN), Tehran, Iran[1]; School of Medicine, Tehran University of Medical Sciences, Tehran, Iran[2]; Research Center for Immunodeficiencies, Children's Medical Center, Tehran University of Medical Sciences, Tehran, Iran[3]

MECHANISM OF ACTION

MECHANISM OF ACTION OF DNA VACCINES

The plasmid DNA is taken up by muscle cells through intramuscular injection and then monocytes express the antigen. Afterward, these antigens are loaded into major histocompatibility complex (MHC)-I molecules on muscle cells, or both MHC-I and MHC-II molecules on antigen-presenting cells (APCs).[1] Also, APCs may capture proteins secreted by transfected cells or contained in transfected apoptotic muscle cells.[1,2]

Because the antigen is expressed in a small amount (nanograms to pictograms), immunogenicity is dependent upon help from CpG islands lying in the sequence of the plasmid backbone that play a role as "built-in" adjuvants, elevating the induction of T cell responses.[3,4] The core CpG motif consists of an unmethylated CpG flanked by two 3′ pyrimidines and two 5′ purines. In vertebrate in comparison to bacterial genomes or viral DNA, CpG sites are relatively rare ($\sim 1\%$). Toll-like receptor (TLR) 9 is expressed in the endosome of APCs including B cells and dendritic cells (DCs). MyD88-dependent signaling cascade is triggered via TLR9 after recognition of CpG, resulting in a proinflammatory response. Moreover, HMGB1 attaches to nucleic acids and is needed for CpG-activated innate immune responses.[5] Thereby, CpG islands augment the DNA vaccines' immunogenicity.[6] Plasmid DNA has a double-stranded sequence that interacts with sensors of cytoplasmic DNA. These sensors, such as TBK-1 and STING, interact with plasmid DNAs and lead to activation of TLR-independent pathways and inducing interferon type 1.[7–10] The DNA methylation does not change its activity, unlike the CpG islands activated TLR9. Rather, a sequence consisting of poly(dAdT)poly(dT-dA) triggers IFN-1 at higher levels compared with poly(dG-dC)poly(dC-dG) sequence, suggesting that the right-handed helical structure of B-DNA is necessary for activation of cellular IFN-1 production.[11] IFN-1 induced via the STING/TBK1 pathway was found to be important for both indirect antigen presentation via muscle cells and direct antigen presentation via DCs.[12]

Vaccines for Cancer Immunotherapy. https://doi.org/10.1016/B978-0-12-814039-0.00010-2

MECHANISM OF ACTION OF RNA VACCINES

The mechanisms of action for RNA vaccines have not been explained completely, however, they likely include some of the same mechanisms used by DNA vaccines for the expression and presentation of encoded antigens resulting in stimulation of immune responses. The RNA is exposed to RNases in the tissue after injection,[13] which can deteriorate the vaccine and restrict uptake of functional RNA by cells. In addition, the presence of 2′-hydroyxl on the ribose sugar prevents the mRNA adopting a stable double β-helix, and allows the macromolecule to be more susceptible to hydrolysis. However, different types of cells are efficient, which use a specific, active and saturable mechanism to internalize the RNA that results in local antigen expression.[14] Uptake is mediated by membrane domains rich in caveolae and lipid rafts, and involves scavenger receptors.[15] A portion of the RNA accumulates in the cytoplasm after internalization and then it is translated into protein. *In situ* antigen production provides a way to imitate pathogen infections and expression of tumor antigens, resulting in effective presentation of antigens by MHC class I and II proteins, and induction of T cell responses in a strategy similar to that provided by viral vectors and DNA vaccines. Alternatively, RNA vaccines can be developed for the efficient production and secretion of extracellular antigens to trigger B cell responses and production of antigen-specific antibody. In addition, the effectiveness of RNA vaccines may be associated with the fact that RNA is known to be a potent stimulator of innate immunity. It has been demonstrated that mRNA activates dendritic cells and monocytes *in vitro* in a MyD88-dependent fashion involving signaling via TLR.[16,17] *In vivo* studies have shown that an mRNA vaccine resulted in the upregulation of different genes involved in cell activation and chemotaxis[18] and also induced TLR7-dependent CD4+ and CD8+ T cell responses, as well as antitumor immunity.[19]

DNA VACCINE

DNA vaccines are modalities that are expressed within the host cells, but they are not capable of replication. They can be produced to imitate the safety and specificity of subunit vaccines. DNA vaccines can trigger immune responses as similar to live-attenuated vaccine types while resulting in no pathogenic infection *in vivo*. Administering DNA vaccines directly into the host lead to expression of the antigenic protein from the host cells. This process induces both antigen-specific antibody and cellular responses.[20–22] Also, applications of DNA vaccines are thought to show safety, stability, and cost-effectiveness, and are simply and rapidly engineered by recent technologies of recombinant DNA.[23]

To deliver DNA vaccines to the target, different approaches including vectors such as viral vectors, naked DNA, liposome and polymer complex; and physical techniques of delivery such as electroporation, ultrasound, and particle-mediated epidermal delivery have been developed.[24] Most of the DNA vaccine studies have used skin or muscle as a target of immunization to deliver DNA vaccines. DNA vaccines are collected and expressed by muscle cells and local APCs in intramuscular injection. For the induction of adaptive immune responses, then local APCs move to the draining lymph nodes.[25,26] The APCs, by cross-presentation of secreted antigen or by transfection of DNA vaccines, are believed to directly present antigen to CD4+ and CD8+ T cells.[22,27,28] According to this, APCs can result in costimulatory signals and cytokines required for stimulation of naïve T cells. In intradermal delivery, DNA can be taken up by dermal dendritic cells and/or Langerhans cells. In this approach, for induction of adaptive immune responses, DNA migrates to the draining lymph nodes.[29]

In particular, DNA vaccines, which stem from bacteria, have unmethylated CpG motifs and stimulate the innate immune responses by interacting with TLR9 expressed on the surface of APCs.[30] This nonspecific activation of APCs likely influences antigen-specific immune responses to DNA vaccines.[23]

Cytotoxic T lymphocytes (CTLs) are expected to be a major effector cell population in cancer immunotherapy targeting neoplastic cells for lysis. Cancer-reactive CTLs kill neoplastic cells through two major killing pathways, the Fas/Fas ligand-mediated and perforin/granzyme B-mediated pathways.[31,32] Cancer cells may become immune-resistant cells regardless of the induction and potentiation of cancer antigen-specific CTL activity and no longer respond to the CTL recognition or attacks. This condition is often seen after long-term immunotherapy or under conditions of immune selection.[33]

Due to the cancer-killing mechanism(s), induced by DNA vaccines and conventional therapy modalities such as chemotherapy, radiation, and surgery, which are all variable, combined therapy utilizing these various modalities might be a choice to restrict the chance of the occurrence of immune-resistant cancer cells, and so lead to improved control of cancer.[23]

CLINICAL TRIALS

To assess whether DNA vaccines have enough efficacy and provide a promising therapeutic approach in treating humans, clinical trials provide the opportunity to test and achieve this final goal. Several clinical trials have revealed the safety of various DNA vaccine platforms since using the first DNA vaccine clinical trial on HIV-1. Plasmid DNA administration is well tolerated by patients with mild or no adverse reactions.[34] The widely evident safety profile of DNA vaccines has resulted in an approval by US Food and Drug Administration for conducting one trial instead of phase I and II trials. The main interest in clinical trials has become demonstrating efficacy due to well-established DNA vaccines safety.[24] In this next discussion we are going to review recent clinical trials and provide a brief summary of the results of trials testing DNA vaccines against cancer.

The first clinical trials of cancer DNA-based vaccines initiated the immunization against the B cell receptor expressed by human B cell lymphomas.[35] It has been shown that intramuscular injection of this DNA vaccine to follicular lymphoma patients in clinical remission after receiving chemotherapy led to promotion of production of anti-idiotypic antibody in 38% of the patients.[35,36] However, preparing a specific idiotypic DNA vaccine for each patient is laborious and costly, which resulted in the development of DNA vaccines coding tumor antigens that are widespreadly present. There are numerous clinical studies utilizing DNA vaccines encoding tumor antigens such as HER-2/neu for breast cancer,[37] E7 for cervical cancer,[38] or gp100 for melanoma.[39] For example, Her2/neu is an oncoprotein playing role in development of malignant phenotype of breast cancer, and it is being targeted by DNA vaccines. In a pilot clinical trial, a vaccine consisting of genetic material of full-length signaling-deficient version of HER2 was administered to patients with metastatic HER2-positive breast cancer. This plasmid DNA was combined with IL-2 and GM-CSF in low doses. It was observed that the vaccine was well tolerated and did not show clinical toxicity or autoimmunity. The vaccine was able to produce long-term antibody responses, and out of the six patients who completed all three cycles of vaccination, two of them survived for more than 4 years after the vaccinations, although no improved T cell responses were observed.[40] In breast cancer, another protein showing overexpression is mammaglobin-A (Mam-A). Preclinical studies illustrated that a DNA vaccine targeting Mam-A can result in CD8+ T cell responses against Mam-A. This resulted in a phase I trial of the Mam-A DNA vaccine in patients with stage IV metastatic breast cancer. Interestingly, it has been observed that Th1 CD4+ T cells were increased. Moreover, the activated CD4+ helper T cells changed from IL-10-expressing phenotype to IFNγ-expressing phenotype and induced preferential destruction of Mam-A-positive breast cancer cells. This finding demonstrates that Mam-A cDNA vaccine is able to show antitumor immunity against breast cancer.[36]

Another example of preclinical and clinical trials of DNA vaccine is in patients with prostate cancer. Preclinical studies have shown that a DNA vaccine targeting prostatic acid phosphatase (PAP) can show PAP-specific CD8+ T cell immune responses. A phase I/II trial was carried out using a DNA vaccine encoding human PAP to be tested in 22 patients with stage D0 prostate cancer. In this study, DNA was injected intradermally six times at 14-day intervals along with the adjuvant GM-CSF (200 mg). 3 of the 22 patients developed PAP-specific IFNγ+ CD8+ T cells and 9 of the patients showed proliferation of PAP-specific CD4+ and CD8+ T cells during 1 year follow up.[41] Several patients were reported to have a decline in the rate of serum PSA rise after treatment, although no PSA values in patients declined by more than 50%.[37] In addition, a plasmid containing the prostate-specific membrane antigen fused to a fragment C domain of tetanus toxin has shown to elicit both CD4+ and CD8+ T cell responses in 3 of 3 recurrent prostate cancer patients.[42]

DELIVERY, IMMUNOGENICITY AND IMMUNE RESPONSE

The major goal of cancer immunotherapy is to introduce different tumor antigens into the host in order to ease the clearance of tumor cells via immune system–mediated strategy. Different kinds of immunotherapy approaches including adoptive T cell therapies, cytokine therapies, antibody therapies, and cancer vaccines have been studied both preclinically and clinically. DNA vaccines, among the different types of cancer vaccines, are recognized as a promising modality to induce robust immune responses. In action, a plasmid DNA that carries antigen genes and other genes of interest is delivered into the tissues of the host to transfect into the cells. These vectors then encode the genes under the control of a mammalian promoter and allow the cells to produce and express them *in vivo* via protein expression machineries of the host.[43] DNA vaccines, depending on their designs and sites of delivery, are demonstrated to be able to trigger innate immune responses, and also able to evoke antigen-specific humoral and cell-mediated immune responses.[36]

To hand over DNA vaccines intradermally, tools like a gene gun can be used, which results in transfection of epidermal keratinocytes and Langerhans cells.[44–46] Langerhans cells are immature dendritic cells located in the skin and play an active role in capturing and processing of antigens. DNA, after delivering to Langerhans cells, results in the expression and processing of antigens, which directly enter the presentation pathway. These transfected Langerhans cells then migrate to the lymph nodes and present antigens to naïve T cells.[47,48] These endogenously produced antigens are presented by MHC class I molecules on Langerhans cells to activate CD8+ CTLs.[45]

DNA vaccines can also be delivered through intramuscular strategy, which leads to the transfection of myocytes. Myocytes are unable of inducing strong specific immune responses because they are not professional APCs, they are effective in expressing the transfected antigens.[27] APCs are gathered to the transfection site. Then APCs through phagocytosis capture antigens generated by transfected cells.[49] These exogenous antigens are then presented by APCs via MHC class II molecules and interact with

CD4+ helper T cells, consequently leading to the activation of a humoral response.[25] Direct transfection of APCs, where direct antigen presentation through MHC class I can prime CD8+ T cells, may be possible.[48,50]

It is important to activate cellular immune responses in eliciting antitumor immunity. Specifically, it has been demonstrated that a potent CD8+ cytotoxic T cell response strongly associates with a good prognosis of tumor control and clearance. DNA vaccines are particularly designed to trigger CD8+ T cell response, and then this response results in intracellular generation of antigens, which activates the MHC class I antigen presentation pathway.[51]

Recent studies of DNA vaccines have demonstrated effectiveness in preclinical studies investigating vaccines against influenza, malaria, and HIV.[52] A regular issue in trials of human DNA vaccine has been their suboptimal immunogenicity when compared with traditional methods of protein-based vaccines.[52] Strategies are required for enhancing DNA vaccine's immunogenicity in the case of human application. Various methods are available to increase the immunogenicity, for example plasmid uptake can be improved by using electroporation, DNA vaccines can benefit from adjuvants by combining a classic adjuvant or adding the encoding gene of new adjuvants, the antigen expression could be optimized through alterations in the codon, and finally using various vectors for priming and boosting lead to increased immunogenicity. However, lacking a surrogate endpoint of vaccine efficacy, the required level of immunogenicity will only be known once a successful outcome study has been achieved. Recent official permission for application of DNA vaccines in veterinary medicine, such as melanoma in dogs, boosts the view that recent technology breakthroughs may be able to produce effective human DNA vaccines. In this regard, more advancement in research can facilitate comprehension of mechanism of action of DNA vaccines to have an overview for optimizing the immunogenicity. In addition, better models are required to accurately evaluate and compare DNA vaccine's potency. A current challenge in the field of DNA vaccine is the poor ability to anticipate human vaccine responses considering mouse immunogenicity data, meaning that there are differences quantitatively between mice and human in DNA vaccine used with adjuvants, which rarely translate to humans.[53]

The genes that are used in vaccination and code for antigens are optimized and are even linked to specific sequences of DNA that encode products behaving as immune-stimulatory or intracellular antigen-targeting molecules. The purpose of this strategy is to enhance the efficacy of translation and immunogenicity of DNA vaccines. Furthermore, cancer DNA vaccines are codelivered with cytokines and/or immune response regulators as adjuvants in order to increase adaptive immune responses. Also, the approach of DNA prime-boost and using xenogeneic antigen are essential strategies to augment adaptive immune responses. These strategies have led to improving the DNA vaccines' immunogenic quality.[23]

ADVERSE EFFECTS

DNA vaccines, which have a good safety profile, are well tolerated by the patients, and without any evident adverse events; possibly the most common adverse reactions are mild to moderate inflammation with related pain, redness, and swelling at the injection site.[52,54] However, there are some reported adverse events related to DNA vaccine in clinical trials. It has been shown that the vector coated with γ-polyglutamic acid (γ-PGA) and composed of spherical nanoparticles is safe and efficient in the lung and spleen after intravenous administration in mice, however, it could cause high liver toxicity and lethality. Applying γ-PGA-coated complex to pUb-M, a melanoma DNA vaccine, noticeably impeded the growth and metastasis of a melanoma cell line, B16–F10 cells.[55] A phase I trial was performed to study safety and immunogenicity of gp100 DNA vaccine on melanoma patients in a cross-over trial manner between mouse and human. In this study, there was not grade 3/4 toxicity in patients. The most common adverse effect was in 12 patients with grade 1/2 (63%, all grade 1) and it was limited to the injection site.[56]

RNA VACCINE

Principles of immunotherapy are based on using the body's own immune system against disease. Cancer vaccines are a type of active immunotherapy with the aim to produce a specific and endogenous immune response to tumor antigens or tumor-associated antigens (TAAs) that are able to target and destroy cancer cells. Tumor antigen itself is one of the components of the cancer vaccine, which can be either one antigen that is expressed by the tumor or a complex mixture of TAAs. Initial investigations demonstrated the role of RNA in immunity by studying the lymphoid tissue RNA extracts of animals injected with tumors that could activate specific immunity when incubated with splenocytes from nonimmunized animals.[57,58] The transferred RNA component that is in charge of the activating immunity is prone to degradation by RNase, and by the use of oligo-dT could be isolated, suggesting that the activity was in the messenger RNA (mRNA) fraction, referred to as "immune RNA."[58]

CLINICAL TRIALS

Until now, only a few trials have been implemented using RNA as a vaccine, and they were restricted to phase I and II clinical trials. Clinical trials have been done using direct administration of mRNA into patients with renal cell carcinoma (RCC),[59] neuroblastoma,[60] pediatric brain cancer,[61] prostate cancer,[62–64] ovarian cancer,[65] lung cancer, breast cancer,[66] and melanoma.[67,68]

Weide et al. implemented two studies on RNA vaccine for metastatic melanoma.[69,70] In the first phase I/II trial *in vitro* transcribed naked whole tumor RNA was administered through the intradermal route to the stages III and IV patients. After 24 h of vaccination, GM-CSF was administered as an adjuvant to the patients. After treatment, melanoma cell line–specific antibodies were detected in 4 of 15 patients, and specific T cell responses were probably induced in 5 patients. The vaccine itself proved to be safe, as only mild and reversible side effects occurred.[70]

In the next phase I/II study, defined RNAs encoding 6 TAAs including survivin, Mage-A1, Mage-A3, gp100, Melan-A, and tyrosinase were administered instead of whole tumor RNA. The escalating doses of RNA were administered to patients with metastatic melanoma. The vaccine was administered more frequent compared with the first trial (12 vaccination within 19 weeks), and the nucleic acids were stabilized with protamine rather than naked RNA. Also, KLH was used as a helper for antigen. Again, 24 h after vaccination, GM-CSF was injected. The analysis of T cell responses led to no consistency between different patients. One patient with augmentation in CD8+ and CD4+ T cells, two patients with negative responses and one patients having an increase and then decrease in the CD8+ and CD4+ T cells were observed. Although the KLH-positive arm resulted in a decline in regulatory T cells (Tregs), the KLH-negative arm showed a decrease in myeloid suppressor cells. In one of seven stage IV patients, clinical response was seen. Again, the study results showed only few achievements by the administered immunotherapy.[69]

DELIVERY, IMMUNOGENICITY AND IMMUNE RESPONSE

The use of mRNA for the induction of antigen-specific immune responses in cancer treatment was reported by Conry et al. in 1995 when they obtained protective antitumor immunity by intramuscular injection of carcinoembryonic antigen (CEA) mRNA.[71] Since then, strategies such as self-replicating RNA, cationic liposomes, and gene gun approach have been introduced for immune intervention utilizing coding mRNA.[72]

Transfection of mRNA into DCs for adoptive transfer was the first mRNA-using to develop vaccine for clinical trials. In 1996, for the first time Boczkowski et al. showed that DCs incubated with mRNA are efficient for presenting the encoded antigen.[73] Promising results in animal models, and the fact that DCs engineered with other compounds had already received regulatory approval for testing in clinical trials, enhanced clinical translation of this concept.[59,63,64,67] Cell transfer is difficult and also lacks major advantages of mRNA such as low cost of goods and simple production process. Direct administration of mRNA was then received great attention. The generation of immune responses with naked but stabilized mRNA was accomplished in multiple mouse models[19,74-77] and encouraged clinical trials.[68,70,78]

To obtain successful clinical results from using vaccine, it is critical to have good immunobioavailability of pertinent epitopes on the surface of professional APCs in the context of MHC molecules *in vivo*. The limited extracellular half-life of naked mRNA due to prompt degradation by ubiquitous RNAs is the major limitation,[13] consequently restricting pharmacologically effective dosing. Liposomes or nanoparticles may protect the RNA or accelerate the uptake but are not easy to produce in clinical grade. According to hypothesis of the defective ribosomal products, which links antigen translation in the cytosol to presentation of peptides on MHC,[79] there is a direct association between the intracellular expression kinetics of mRNA encoded antigen and the extent of the elicited immune response.[77]

Intracellular stability and translational efficacy of mRNA can be increased significantly by structural alterations of the 5′ cap analog, the untranslated region (UTRs), and the poly(A) tail. A 700-fold increase of luciferase expression was reported via transfer of an mRNA with an antireverse cap analog and a 100 bp poly(A) tail as compared to 65 bp poly(A) tail and a standard cap (m7GpppG); 120 bp length has been demonstrated to be optimal for a poly(A) tail and together with a 3′βglobin UTR increases transcript stability by two orders of magnitude. Novel cap analogs contributed to extremely higher frequencies of antigen-specific T cells. Recently, an internal ribosomal entry site as translation initiator was shown to increase luciferase expression in human DCs.[80]

ADVERSE EFFECTS

RNA-based cancer vaccines seem to be safe, well tolerated by patients, and without any severe and noticeable adverse events in clinical trial.[81,82] Schmidt et al. performed a study to test naked tumor RNA coding for the TAAs survivin, Her-2/neu, MAGE-1, MUC1, CEA, and telomerase in 30 patients with RCC. The patients were divided into two groups; group A received intradermal injection on days 1, 14, 28, and 42, and group B received vaccinations on days 0–3, 7–10, 28, and 42 following monthly injections. ELISPOT assays were used to evaluate the generation of antigen-specific immune responses *in vitro*. Both CD4+ and CD8+ T cell responses were developed for diverse antigens. Moreover, seven clinical responses were observed among the patients. No severe negative side effects were reported, displaying the safety profile of the approach.[82]

ADVANTAGES AND DISADVANTAGES

ADVANTAGES AND DISADVANTAGES OF DNA VACCINES

There are several advantages of *in vivo* use of DNA vaccine to produce antigens for cancer immunotherapy. Technology of molecular recombinant DNA makes it possible to design DNA vectors in order to encode a wide range of antigens and immunomodulatory molecules.[83] Also, the DNA plasmids have intrinsic abilities to trigger innate immune responses.[36] By using different routes of application and optimizing the antigen processing pathways, DNA vaccines can preferentially induce the activation of either Th1 cells or Th2 cells and polarize the resulting immune response into being either cellular or humoral based. Also, DNA vaccines can evoke CD8+ CTL-mediated immune responses with significant role for tumor killing.[51] Transfection of DNA plasmid into host cells permits for steady expression and supply of antigens. DNA vaccines manifest the potential to be a mass application approach for cancer immunotherapy. Surprisingly, they can be engineered easily and produced rapidly in large quantities. Vectors of DNA are very stable and can be transported and stored easily. It has also been demonstrated that using DNA vaccines in both animal models and human clinical trials is safe.[38,56,84] The vaccines do not lead to pathogenic infection unlike with the live attenuated bacterial or viral vaccines. Moreover, the body do not produce antivector neutralizing antibody; therefore, several doses of vaccines can administered. [85] Different ways of DNA vaccine delivery have showed only limited, tolerable inconvenience with no substantial adverse effects. Advantages such as safety, simple manufacturing, and inexpensive production make DNA vaccines an attractive choice.[24]

In spite of the advantages of DNA vaccines, these cancer vaccines still have disadvantages, such as relatively weak immunogenicity, which inhibits providing the desired clinical success. Furthermore, *in vivo* administration of naked DNA does not lead to distribution from cell to cell simply. Capturing the expressed antigens by APCs and triggering potent immune responses are not performed simply. Therefore, effective strategies need to be developed in order to help augment DNA vaccine potency[24] (Fig. 10.1).

ADVANTAGES OF RNA VACCINES

As part of the cellular mechanism, RNA vaccines produce different peptides in the patient. Subsequently, some of the generated peptides bind to the patient's HLA molecules. Therefore, there is no restriction to prior immunogenic peptide identification or knowledge of the patient's HLA type in the RNA-based vaccine-treated patients. Many reports have shown the advantages of inducing CD4+ and CD8+ T cells for producing an effective and long-lasting immune response against tumors. However, production and purification of protein is an overwhelming process, limiting its attractiveness as a source of antigen. One strategy is that DNA should be delivered to the cytoplasm of the cell and then must travel to the nucleus to be transcribed into mRNA. Then, the mRNA is translated into a protein that is then subjected to mechanism of MHC class I processing in the cells to produce the relevant MHC-binding peptides. Delivering the DNA to the nucleus is not always efficient, although cells have

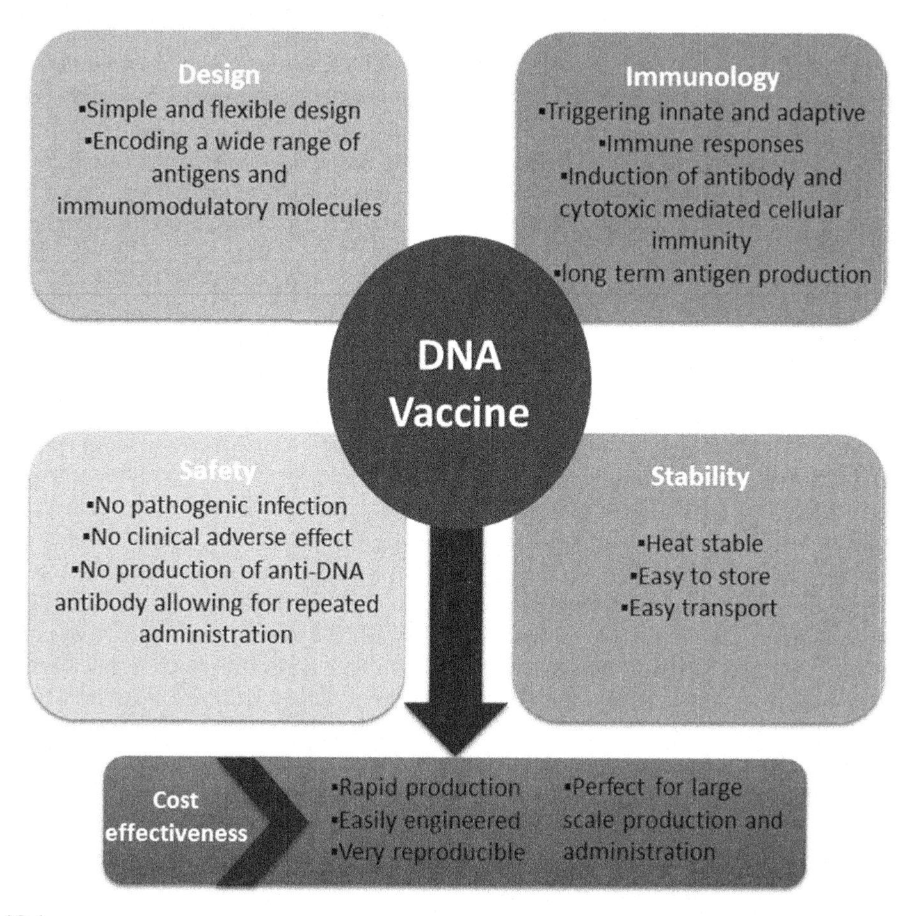

FIGURE 10.1

Advantages of DNA vaccine.

been transfected with DNA by various methods, including electroporation and cationic lipofection.[86] However, mRNA transferred to the cytoplasm is readily translated into proteins. A second potential problem correlated with DNA use is that the DNA can integrate into the host-cell genome, and may possibly defect the gene structure involved in an important signaling pathway and neoplasm process. This problem is largely theoretical, and eradicated in RNA-based vaccines. Van Tendeloo et al. showed another potential reason to select mRNA and why it is better than DNA by comparing Langerhans cells (epidermal DCs) stemmed from CD34+ precursor, electroporated with Melan-A-encoding DNA or mRNA for their ability to stimulate interferon-γ (IFNγ) production from a Melan-A-specific CTL clone.[86] Although both the DNA-transfected and mRNA-transfected cells triggered the Melan-A CTL clone, cells electroporated with DNA encoding a protein other than Melan-A caused nonspecific stimulation. But when mRNA was utilized, only cells electroporated with mRNA encoding Melan-A caused interferon-γ release from the CTL clone, showing the improved specificity of this method.

Generating large quantities of mRNA, which transcribes under good manufacturing practices *in vitro*, is an inexpensive and straightforward process if TAA expressed by a particular cancer is known. In this regard, it allows a one-time cloning of the appropriate cDNA into a vector that encodes a bacteriophage promoter and a poly(A) tail. For mRNA production, the cDNA-containing plasmid is linearized by restriction enzyme and utilized as a template for *in vitro* transcription in a reaction that contains bacteriophage RNA polymerase, ribonucleotide triphosphates, and buffers. After transcription, the plasmid template is digested with DNase and the mRNA is cleaned up for subsequent use.

OPTIMIZATION

DNA VACCINE OPTIMIZATION

Optimization of nonviral DNA vaccines is the most important strategy for their expression. Various approaches as part of expression cassette,[54] such as promoters,[87] transgenes encoding specific antigens, the 3′ UTR and polyadenylation signals, and vector backbones[88] can be used successfully for optimization of DNA vaccines. These modifications are important for increased expression.

Promoters of vectors can be modified by strengthening the potency of promoters in order to increase the DNA-based vaccines' efficacy. The strength of the promoter is necessary for a potent immune response. The ubiquitous promoters can be used to result in the high expression of proteins in most cell types. One of the most potent promoters known is the cytomegalovirus (CMV) immediate-early promoter.[89] But, the CMV promoter has its own limitations. For instance, it has been demonstrated that transgene expression from the CMV promoter promptly declined after intravenous and intramuscular injection of adenovectors compared with other promoters.[90] Some vectors and promoters are more unique to cancer cells. Alphavirus vector systems are viruses with positive strand RNA that are being generated for the potential treatment of progressive and refractory tumors.[91] Deregulation of c-myc promoter inhibit the terminal differentiation of cells in cancer[92]; murine leukemia virus (MLV)-derived replication-competent retrovirus vectors show very specific features to transfect tumor cells.[93] Survivin is an inhibitor of apoptosis that is expressed in cancer cells but not in normal cells. It has been revealed that survivin could be a potential candidate promoter. *In vivo* studies of this promoter has been demonstrated to be 200-fold more cancer specific than the CMV promoter as it was expressed in 81% of the non-small-cell lung cancers tested but was absent in normal lung tissues.[94]

Expression of the HPV oncogenic proteins E6 and E7 via HPV type-16 infections are needed for cervical cancer development, progression, and preservation.[95] It has been shown that the E7 protein triggers immune responses in cervical cancer patients, thus it has become the focus target for immunotherapy. E7 codon optimization by replacing infrequent-use codons with high priority human codons led to protein levels that were increased six-to ninefold compared with wild-type HPV16.[96,97]

RNA VACCINE OPTIMIZATION

In the immune system, regulatory CD4+ CD25+ T cells are essential for self-tolerance and have suppressor functions. This type of cell controls immune responses and reduces the risk of T cell responses being harmful to the body. Increased numbers of Tregs are associated with a reduced survival in patients with tumor. Therefore, the depletion of Tregs could strengthen the induced immune responses and prolong the life of patients. Dannull et al. evaluated patients with RCC by RNA-transfected DCs, and undertook a strategy to decrease the number of regulatory T cells. They utilized the recombinant IL-2 diphtheria toxin conjugate DAB_{389}IL-2 (ONTAK), which selectively targeted and decreased CD25-positive regulatory T cells.[65]

It is also feasible to stimulate mechanisms that activate the immune system in addition to inhibition of undesired immune effects. For example, nonmethylated CpGs are compounds with enhancing qualities.[98] These synthetic oligodeoxynucleotides are homologues to DNA of bacteria or virus and stimulate TLR9. TLR9 is located in the endosomal compartments of macrophages and dendritic cells, and triggers APCs by binding to PAMPS. For example, synthetic oligodeoxynucleotide PF-3512676, elevates the induced immune response and prevents induction of immune tolerance.[99]

Combination of RNA vaccination and tyrosine kinase inhibitors (TKIs) is considered as another strategy to improve treatment efficacy. The TKIs, such as sorafenib and sunitinib, impede intracellular signaling pathways that are responsible for proliferation and angiogenesis. Sorafenib is administered in the therapy of RCC and hepatocellular carcinoma. Sunitinib is used in gastrointestinal tumor and RCC treatment. Recent studies showed that pretreatment of mouse with sorafenib decreased the antigen-specific T cell induction, however, sunitinib had no such effect.[100] Sunitinib had no impact on T cell proliferation and phenotype of human monocyte-derived DCs, while sorafenib impeded maturation processes in DCs and the stimulation of T cells. These findings illustrate that sunitinib might be a potential candidate for combinational therapy with RNA vaccinations.[101]

REFERENCES

1. Kutzler MA, Weiner DB. Developing DNA vaccines that call to dendritic cells. *J Clin Investig* 2004;**114**(9): 1241–4.
2. Cho JH, Youn JW, Sung YC. Cross-priming as a predominant mechanism for inducing CD8+ T cell responses in gene gun DNA immunization. *J Immunol* 2001;**167**(10):5549–57.
3. Dalpke A, Zimmermann S, Heeg K. CpG-oligonucleotides in vaccination: signaling and mechanisms of action. *Immunobiology* 2001;**204**(5):667–76.
4. Krieg AM, et al. CpG motifs in bacterial DNA trigger direct B-cell activation. *Nature* 1995;**374**(6522):546.
5. Yanai H, Ban T, Taniguchi T. Essential role of high-mobility group box proteins in nucleic acid-mediated innate immune responses. *J Intern Med* 2011;**270**(4):301–8.
6. Klinman DM, Yamshchikov G, Ishigatsubo Y. Contribution of CpG motifs to the immunogenicity of DNA vaccines. *J Immunol* 1997;**158**(8):3635–9.
7. Ishii KJ, et al. A Toll-like receptor–independent antiviral response induced by double-stranded B-form DNA. *Nat Immunol* 2006;**7**(1):40.
8. Coban C, et al. Molecular and cellular mechanisms of DNA vaccines. *Hum Vaccines* 2008;**4**(6):453–7.
9. Ishii KJ, et al. TANK-binding kinase-1 delineates innate and adaptive immune responses to DNA vaccines. *Nature* 2008;**451**(7179):725.
10. Ishikawa H, Ma Z, Barber GN. STING regulates intracellular DNA-mediated, type I interferon-dependent innate immunity. *Nature* 2009;**461**(7265):788.
11. Koyama S, et al. Innate immune control of nucleic acid-based vaccine immunogenicity. *Expert Rev Vaccines* 2009;**8**(8):1099–107.
12. Coban C, et al. Novel strategies to improve DNA vaccine immunogenicity. *Curr Gene Ther* 2011;**11**(6): 479–84.

13. Probst J, et al. Characterization of the ribonuclease activity on the skin surface. *Genet Vaccine Ther* 2006;**4**(1):4.
14. Probst J, et al. Spontaneous cellular uptake of exogenous messenger RNA in vivo is nucleic acid-specific, saturable and ion dependent. *Gene Therapy* 2007;**14**(15):1175.
15. Lorenz C, et al. Protein expression from exogenous mRNA: uptake by receptor-mediated endocytosis and trafficking via the lysosomal pathway. *RNA Biology* 2011;**8**(4):627–36.
16. Scheel B, et al. Toll-like receptor-dependent activation of several human blood cell types by protamine-condensed mRNA. *Eur J Immunol* 2005;**35**(5):1557–66.
17. Karikó K, et al. mRNA is an endogenous ligand for Toll-like receptor 3. *J Biol Chem* 2004;**279**(13):12542–50.
18. Fotin-Mleczek M, et al. Highly potent mRNA based cancer vaccines represent an attractive platform for combination therapies supporting an improved therapeutic effect. *J Gene Med* 2012;**14**(6):428–39.
19. Fotin-Mleczek M, et al. Messenger RNA-based vaccines with dual activity induce balanced TLR-7 dependent adaptive immune responses and provide antitumor activity. *J Immunother* 2011;**34**(1):1–15.
20. Kuhöber A, et al. DNA immunization induces antibody and cytotoxic T cell responses to hepatitis B core antigen in H-2b mice. *J Immunol* 1996;**156**(10):3687–95.
21. Wang B, et al. Gene inoculation generates immune responses against human immunodeficiency virus type 1. *Proc Natl Acad Sci USA* 1993;**90**(9):4156–60.
22. Corr M, et al. Gene vaccination with naked plasmid DNA: mechanism of CTL priming. *J Exp Med* 1996;**184**(4):1555–60.
23. Lee S-H, Danishmalik SN, Sin J-I. DNA vaccines, electroporation and their applications in cancer treatment. *Hum Vaccines Immunother* 2015;**11**(8):1889–900.
24. Yang B, et al. DNA vaccine for cancer immunotherapy. *Hum Vaccines Immunother* 2014;**10**(11):3153–64.
25. Casares S, et al. Antigen presentation by dendritic cells after immunization with DNA encoding a major histocompatibility complex class II–restricted viral epitope. *J Exp Med* 1997;**186**(9):1481–6.
26. Chattergoon MA, et al. Specific immune induction following DNA-based immunization through in vivo transfection and activation of macrophages/antigen-presenting cells. *J Immunol* 1998;**160**(12):5707–18.
27. Fu T-M, et al. Priming of cytotoxic T lymphocytes by DNA vaccines: requirement for professional antigen presenting cells and evidence for antigen transfer from myocytes. *Mol Med* 1997;**3**(6):362.
28. Iwasaki A, et al. The dominant role of bone marrow-derived cells in CTL induction following plasmid DNA immunization at different sites. *J Immunol* 1997;**159**(1):11–4.
29. Condon C, et al. DNA–based immunization by in vivo transfection of dendritic cells. *Nat Med* 1996;**2**(10):1122.
30. Liu MA, Ulmer JB. Human clinical trials of plasmid DNA vaccines. *Adv Genet* 2005;**55**:25–40.
31. Atkinson EA, Bleackley RC. Mechanisms of lysis by cytotoxic T cells. *Critic Rev Immunol* 1995;**15**(3–4).
32. Groscurth P, Filgueira L. Killing mechanisms of cytotoxic T lymphocytes. *Physiology* 1998;**13**(1):17–21.
33. Igney FH, Krammer PH. Immune escape of tumors: apoptosis resistance and tumor counterattack. *J Leukoc Biol* 2002;**71**(6):907–20.
34. Ledwith BJ, et al. Plasmid DNA vaccines: investigation of integration into host cellular DNA following intramuscular injection in mice. *Intervirology* 2000;**43**(4–6):258–72.
35. Hawkins RE, et al. Idiotypic vaccination against human B-cell lymphoma. Rescue of variable region gene sequences from biopsy material for assembly as single-chain Fv personal vaccines. *Blood* 1994;**83**(11):3279–88.
36. Rice J, Ottensmeier CH, Stevenson FK. DNA vaccines: precision tools for activating effective immunity against cancer. *Nat Rev Cancer* 2008;**8**(2):108.
37. Disis ML, et al. Effect of dose on immune response in patients vaccinated with an her-2/neu intracellular domain protein–based vaccine. *J Clin Oncol* 2004;**22**(10):1916–25.

38. Trimble CL, et al. A phase I trial of a human papillomavirus DNA vaccine for HPV16+ cervical intraepithelial neoplasia 2/3. *Clin Cancer Res* 2009;**15**(1):361–7.
39. Perales M-A, et al. Phase I/II study of GM-CSF DNA as an adjuvant for a multipeptide cancer vaccine in patients with advanced melanoma. *Mol Ther* 2008;**16**(12):2022–9.
40. Liu J, et al. Tumor-associated macrophages recruit CCR6+ regulatory T cells and promote the development of colorectal cancer via enhancing CCL20 production in mice. *PLoS One* 2011;**6**(4):e19495.
41. Timmerman JM, et al. Immunogenicity of a plasmid DNA vaccine encoding chimeric idiotype in patients with B-cell lymphoma. *Cancer Res* 2002;**62**(20):5845–52.
42. Maldonado L, et al. Intramuscular therapeutic vaccination targeting HPV16 induces T cell responses that localize in mucosal lesions. *Sci Transl Med* 2014;**6**(221):221ra13.
43. Webster RG, Robinson HL. DNA vaccines: a review of developments. *BioDrugs* 1997;**8**(4):273–92.
44. Fuller DH, Loudon P, Schmaljohn C. Preclinical and clinical progress of particle-mediated DNA vaccines for infectious diseases. *Methods* 2006;**40**(1):86–97.
45. Trimble C, et al. Comparison of the CD8+ T cell responses and antitumor effects generated by DNA vaccine administered through gene gun, biojector, and syringe. *Vaccine* 2003;**21**(25–26):4036–42.
46. Lauterbach H, et al. Insufficient APC capacities of dendritic cells in gene gun-mediated DNA vaccination. *J Immunol* 2006;**176**(8):4600–7.
47. Porgador A, et al. Predominant role for directly transfected dendritic cells in antigen presentation to CD8+ T cells after gene gun immunization. *J Exp Med* 1998;**188**(6):1075–82.
48. Akbari O, et al. DNA vaccination: transfection and activation of dendritic cells as key events for immunity. *J Exp Med* 1999;**189**(1):169–78.
49. Dupuis M, et al. Distribution of DNA vaccines determines their immunogenicity after intramuscular injection in mice. *J Immunol* 2000;**165**(5):2850–8.
50. Ulmer J, Otten G. Priming of CTL responses by DNA vaccines: direct transfection of antigen presenting cells versus cross-priming. *Dev Biologicals* 2000;**104**:9–14.
51. Anderson RJ, Schneider J. Plasmid DNA and viral vector-based vaccines for the treatment of cancer. *Vaccine* 2007;**25**:B24–34.
52. Klinman DM, et al. FDA guidance on prophylactic DNA vaccines: analysis and recommendations. *Vaccine* 2010;**28**(16):2801–5.
53. Saade F, Petrovsky N. Technologies for enhanced efficacy of DNA vaccines. *Expert Rev Vaccine* 2012;**11**(2):189–209.
54. Jechlinger W. Optimization and delivery of plasmid DNA for vaccination. *Expert Rev Vaccine* 2006;**5**(6):803–25.
55. Kurosaki T, et al. Secure splenic delivery of plasmid DNA and its application to DNA vaccine. *Biol Pharm Bull* 2013;**36**(11):1800–6.
56. Yuan J, et al. Safety and immunogenicity of a human and mouse gp100 DNA vaccine in a phase I trial of patients with melanoma. *Cancer Immun* 2009;**9**:5.
57. Fishman M, Adler FL. Antibody formation initiated in vitro: II. Antibody synthesis in X-irradiated recipients of diffusion chambers containing nucleic acid derived from macrophages incubated with antigen. *J Exp Med* 1963;**117**(4):595–602.
58. Pilch YH, Ramming KP. Transfer of tumor immunity with ribonucleic acid. *Cancer* 1970;**26**(3):630–7.
59. Su Z, et al. Immunological and clinical responses in metastatic renal cancer patients vaccinated with tumor RNA-transfected dendritic cells. *Cancer Res* 2003;**63**(9):2127–33.
60. Caruso DA, et al. Results of a phase I study utilizing monocyte-derived dendritic cells pulsed with tumor RNA in children with stage 4 neuroblastoma. *Cancer* 2005;**103**(6):1280–91.
61. Caruso DA, et al. Results of a phase 1 study utilizing monocyte-derived dendritic cells pulsed with tumor RNA in children and young adults with brain cancer. *NeuroOncology* 2004;**6**(3):236–46.

62. Mu L, et al. Immunotherapy with allotumour mRNA-transfected dendritic cells in androgen-resistant prostate cancer patients. *Br J Cancer* 2005;**93**(7):749.
63. Heiser A, et al. Autologous dendritic cells transfected with prostate-specific antigen RNA stimulate CTL responses against metastatic prostate tumors. *J Clin Investig* 2002;**109**(3):409–17.
64. Su Z, et al. Telomerase mRNA-transfected dendritic cells stimulate antigen-specific CD8+ and CD4+ T cell responses in patients with metastatic prostate cancer. *J Immunol* 2005;**174**(6):3798–807.
65. Dannull J, et al. Enhancement of vaccine-mediated antitumor immunity in cancer patients after depletion of regulatory T cells. *J Clin Investig* 2005;**115**(12):3623–33.
66. Morse MA, et al. Immunotherapy with autologous, human dendritic cells transfected with carcinoembryonic antigen mRNA. *Cancer Investig* 2003;**21**(3):341–9.
67. Kyte J, et al. Phase I/II trial of melanoma therapy with dendritic cells transfected with autologous tumor-mRNA. *Cancer Gene Ther* 2006;**13**(10):905.
68. Weide B, et al. Direct injection of protamine-protected mRNA: results of a phase 1/2 vaccination trial in metastatic melanoma patients. *J Immunother* 2009;**32**(5):498–507.
69. Weide B, et al. Plasmid DNA-and messenger RNA-based anti-cancer vaccination. *Immunol Lett* 2008;**115**(1):33–42.
70. Weide B, et al. Results of the first phase I/II clinical vaccination trial with direct injection of mRNA. *J Immunother* 2008;**31**(2):180–8.
71. Conry RM, et al. Characterization of a messenger RNA polynucleotide vaccine vector. *Cancer Res* 1995;**55**(7):1397–400.
72. Pascolo S. Messenger RNA-based vaccines. *Expert Opin Biol Ther* 2004;**4**(8):1285–94.
73. Boczkowski D, et al. Dendritic cells pulsed with RNA are potent antigen-presenting cells in vitro and in vivo. *J Exp Med* 1996;**184**(2):465–72.
74. Hoerr I, et al. In vivo application of RNA leads to induction of specific cytotoxic T lymphocytes and antibodies. *Eur J Immunol* 2000;**30**(1):1–7.
75. Carralot J-P, et al. Polarization of immunity induced by direct injection of naked sequence-stabilized mRNA vaccines. *Cell Mol Life Sci* 2004;**61**(18):2418–24.
76. Granstein RD, Ding W, Ozawa H. Induction of anti-tumor immunity with epidermal cells pulsed with tumor-derived RNA or intradermal administration of RNA. *J Investig Dermatol* 2000;**114**(4):632–6.
77. Kreiter S, et al. Intranodal vaccination with naked antigen-encoding RNA elicits potent prophylactic and therapeutic antitumoral immunity. *Cancer Res* 2010;**70**(22):9031–40.
78. Rittig SM, et al. Intradermal vaccinations with RNA coding for TAA generate CD8+ and CD4+ immune responses and induce clinical benefit in vaccinated patients. *Mol Ther* 2011;**19**(5):990–9.
79. Princiotta MF, et al. Quantitating protein synthesis, degradation, and endogenous antigen processing. *Immunity* 2003;**18**(3):343–54.
80. Kreiter S, et al. Tumor vaccination using messenger RNA: prospects of a future therapy. *Curr Opin Immunol* 2011;**23**(3):399–406.
81. Li J, et al. Messenger RNA vaccine based on recombinant MS2 virus-like particles against prostate cancer. *Int J Cancer* 2014;**134**(7):1683–94.
82. Schmidt S, et al. Vaccinations with RNA coding for tumor associated antigens in advanced RCC patients—a phase I/II study. *J Clin Oncol* 2008;**26**(15 Suppl.):3017.
83. Fioretti D, et al. DNA vaccines: developing new strategies against cancer. *BioMed Res Int* 2010;**2010**.
84. Staff C, et al. A Phase I safety study of plasmid DNA immunization targeting carcinoembryonic antigen in colorectal cancer patients. *Vaccine* 2011;**29**(39):6817–22.
85. Smith H. Regulation and review of DNA vaccine products. *Dev Biologicals* 2000;**104**:57–62.

86. Van Tendeloo VF, et al. Highly efficient gene delivery by mRNA electroporation in human hematopoietic cells: superiority to lipofection and passive pulsing of mRNA and to electroporation of plasmid cDNA for tumor antigen loading of dendritic cells. *Blood* 2001;**98**(1):49–56.
87. Li X, et al. Synthetic muscle promoters: activities exceeding naturally occurring regulatory sequences. *Nat Biotechnol* 1999;**17**(3):241.
88. Wagner E. Advances in cancer gene therapy: tumor-targeted delivery of therapeutic pDNA, siRNA, and dsRNA nucleic acids. *J BU Oncology* 2007;**12**:S77–82.
89. Wright A, et al. Diverse plasmid DNA vectors by directed molecular evolution of cytomegalovirus promoters. *Hum Gene Ther* 2005;**16**(7):881–92.
90. Chen P, et al. Promoters influence the kinetics of transgene expression following adenovector gene delivery. *J Gene Med* 2008;**10**(2):123–31.
91. Kelly BJ, Fleeton MN, Atkins GJ. Potential of alphavirus vectors in the treatment of advanced solid tumors. *Recent Patents on Anti-Cancer Drug Discovery* 2007;**2**(2):159–66.
92. Wierstra I, Alves J. The c-myc promoter: still mysterY and challenge. *Adv Cancer Res* 2008;**99**:113–333.
93. Tai C-K, Kasahara N. Replication-competent retrovirus vectors for cancer gene therapy. *Front Biosci* 2008; **13**:3083–95.
94. Chen J-S, et al. Cancer-specific activation of the survivin promoter and its potential use in gene therapy. *Cancer Gene Ther* 2004;**11**(11):740.
95. Narisawa-Saito M, Kiyono T. Basic mechanisms of high-risk human papillomavirus-induced carcinogenesis: roles of E6 and E7 proteins. *Cancer Sci* 2007;**98**(10):1505–11.
96. Disbrow GL, et al. Codon optimization of the HPV-16 E5 gene enhances protein expression. *Virology* 2003; **311**(1):105–14.
97. Cid-Arregui A, Juárez V, zur Hausen H. A synthetic E7 gene of human papillomavirus type 16 that yields enhanced expression of the protein in mammalian cells and is useful for DNA immunization studies. *J Virol* 2003;**77**(8):4928–37.
98. Vollmer J, et al. Characterization of three CpG oligodeoxynucleotide classes with distinct immunostimulatory activities. *Eur J Immunol* 2004;**34**(1):251–62.
99. Molenkamp BG, et al. Local administration of PF-3512676 CpG-B instigates tumor-specific CD8+ T-cell reactivity in melanoma patients. *Clin Cancer Res* 2008;**14**(14):4532–42.
100. Hipp MM, et al. Sorafenib, but not sunitinib, affects function of dendritic cells and induction of primary immune responses. *Blood* 2008;**111**(12):5610–20.
101. Bringmann A, et al. RNA vaccines in cancer treatment. *BioMed Res Int* 2010;**2010**.

CHAPTER

CANDIDATE CANCERS FOR VACCINATION

11

Mahsa Keshavarz-Fathi[1,2,3], **Nima Rezaei**[3,4,5]

School of Medicine, Tehran University of Medical Sciences, Tehran, Iran[1]*; Cancer Immunology Project (CIP), Universal Scientific Education and Research Network (USERN), Tehran, Iran*[2]*; Research Center for Immunodeficiencies, Children's Medical Center, Tehran University of Medical Sciences, Tehran, Iran*[3]*; Department of Immunology, School of Medicine, Tehran University of Medical Sciences, Tehran, Iran*[4]*; Network of Immunity in Infection, Malignancy and Autoimmunity (NIIMA), Universal Scientific Education and Research Network (USERN), Tehran, Iran*[5]

Immunotherapy and cancer vaccines have been evaluated for many types of cancer. Most clinical trials that tested cancer vaccines were conducted in patients with progressive or metastatic disease who had previously been treated with different modalities.[1] However, it has been demonstrated that patients at an early stage of cancer and with fewer prior chemotherapeutic agents with a longer time passed from the chemotherapy will show better response to vaccines.[2,3]

The immunosuppressive setting of a tumor is the other determinant of response to vaccines. Because the immunosuppressive microenvironment is more expected in high-grade and large tumors, vaccines might be less effective compared with chemotherapy in this setting. However, the safety profile of vaccines is superior to that of chemotherapy.[1,4]

Until recently, vaccines were assessed for the treatment of many types of cancer. There are clinical trials involving patients with different types of cancer, such as melanoma, prostate, lung, colorectal, breast, renal cell carcinoma, pancreatic, ovarian carcinoma, and hematologic malignancies.[5–12]

Many attempts have been made to use cancer vaccines in patients with melanoma because of the higher rate of mutation in this malignancy.[13] However, prostate cancer has a number of special features that made it a prototype for treatment with cancer vaccines: (1) Prostate cancer is an indolent disease in most patients and is in concordance with the sufficient time, which is usually needed to develop an immune response capable of destroying tumor cells. (2) Various tumor-associated and tumor-specific antigens are available on the surface of prostate cancer, which are potential targets for specific immunotherapy including vaccines. (3) The serum prostate-specific antigen (PSA) level is a good indicator for prostate cancer follow-up and response to treatment. (4) The Halabi nomogram can predict the survival of metastatic hormone-refractory prostate cancer patients who receive the standard of care.[1,14]

VACCINES FOR PROSTATE CANCER

Sipuleucel-T (Provenge) is a cancer vaccine approved by the US Food and Drug Administration to treat men with hormone-refractory prostate cancer. It is a personalized vaccine including

Vaccines for Cancer Immunotherapy. https://doi.org/10.1016/B978-0-12-814039-0.00011-4

antigen presenting cells extracted from the peripheral blood mononuclear cells of the patient and incubated with a protein consisting of a tumor antigen and an adjuvant, i.e., granulocyte macrophage colony-stimulating factor (GM-CSF) and prostatic acid phosphatase.[15,16] It was evaluated in three clinical trials that showed its superiority to placebo in men who were resistant to hormone therapy.[17] Sipuleucel-T showed a 4.1- to 4.3-month increase in the overall survival (OS) and a 22% decrease in the risk for death.[18] Patients with a lower tumor burden benefit more from sipuleucel-T because 13 months improvement in OS was gained in patients with lower PSA levels whereas 2.8 months improvement in OS was seen in those with a higher PSA level.[19] Sipuleucel-T is being evaluated in combination with other therapies including Ipilimumab and the approved androgen inhibitors and radiotherapeutic agents.[17]

PROSTVAC-VF (TRICOM) is a genetic-based vaccine consisting of a recombinant poxvirus vector for the first vaccination and a recombinant fowlpox vector for the booster doses, which are engineered to express PSA and three costimulatory molecules: B7.1 (CD80), lymphocyte function-associated antigen 3 (LFA-3), and intracellular adhesion molecule 1 (ICAM-1).[20] PSA reduction and PSA stabilization for 11–21 months were reported in phase I clinical trials with PROSTVAC-VF.[21] In a phase II trial, the 3-year OS was reported 30% in patients treated with PROSTVAC, compared with 17% in the control arm.[22] A phase III trial was completed in patients randomized to three arms: PROSTVAC-V/F coadministered with GM-CSF, PROSTVAC-V/F coadministered with GM-CSF placebo, and double placebo (NCT01322490).

ProstAtak (AdV-tk) is a gene-mediated cytotoxic immunotherapy using an adenoviral vector to a cytotoxic gene, i.e., a Herpes virus thymidine-kinase gene (AdV-tk) to the tumor cells. After administration of ProstAtak, a systemic antiherpes prodrug, valacyclovir, is administered. ProstAtak is used combined with standard therapies such as surgery and radiation.[23] It is being tested in combination with radiation therapy in an ongoing phase III trial (NCT01436968).

VACCINES FOR MELANOMA

Much time and effort have been put into treating melanoma with cancer vaccines because it has the highest mutation rate among different cancers and contains a variety of tumor antigens.[13] Although, no cancer vaccine is approved for treating melanoma yet, it is an attractive candidate for being targeted with individualized vaccines.

A review by Rosenberg et al.[24] was published in 2004, reporting all of the trials that they carried out to evaluate cancer vaccines in melanoma. Various adjuvants and vectors were used with a broad coverage of tumor antigens; however, the objective response rate was low. Afterward, the gp100 peptide vaccine with the adjuvant Montanide ISA-51 and systemic interleukin-2 (IL-2) was administered to patients with advanced melanoma and compared with the efficacy of IL-2 alone. The OS was improved with the vaccine (17.8 months for the vaccine arm and 11.1 months for IL-2 alone; $P = 0.06$).[25] A trial with gp100 vaccine combined with ipilimumab, an anti- CTLA-4 antibody, was conducted, and resulted in a 3-year survival rate of 25% in the ipilimumab arm and 15% in the combination arm.[26]

Oncophage is a personalized vaccine consisting of tumor-derived heat shock proteins, which contains peptides of the tumor. Oncophage was evaluated in a trial in patients with stage IV melanoma, with the aim of comparing its efficacy with the physician's choice. The OS was similar between the two arms.[27]

A dendritic cell (DC)-based vaccine loaded *ex vivo* with major histocompatibility complex class I and II-restricted peptides was also compared with standard dacarbazine in a phase III trial in metastatic melanoma. Similar objective responses were derived for the two arms.[28]

The lack of vaccine superiority over other treatments shifted the focus of studies to using neoantigens for vaccination to treat melanoma. Clinical trials evaluated peptide-based and RNA-based vaccines to enhance disease-free survival (DFS) in patients at high risk for melanoma recurrence. Remarkable clinical responses resulted in these phase I trials, which were grounds for applying a personalized vaccine approach on a greater scale.[29,30]

In one of these trials, a cancer vaccine with up to 20 neoantigens of melanoma led to more than 2 years DFS in four of six patients. The other two patients with melanoma, which had progressed, had regression after administration of an Anti programmed cell death-1 (PD-1) antibody.[31] In the other trial, patients with melanoma were treated with a poly-neo-epitope RNA vaccine; all patients showed an increase in progression-free survival. Patients without symptomatic melanoma before vaccination remained recurrence-free for 12–23 months; three out of five of those who had progressed before vaccination showed a complete response. The other two patients showed a partial response and stable disease.[32]

VACCINES FOR LUNG CANCER

One of the known peptide vaccines examined in lung cancer is the melanoma-associated antigen-A3 (MAGE-A3) vaccine. MAGE-A3 is a cancer-testes antigen normally expressed only by male germ cells. It is also expressed by tumor cells, for instance, it is expressed in 30–50% of patients with non–small cell lung cancer (NSCLC).[33,34] This vaccine was evaluated in a phase II trial. Patients with resected NSCLC were randomized into two arms: MAGE-A3 plus adjuvant or placebo. After a median follow-up of 28 months, hazard ratios of DFS and OS were 0.73 (95% confidence interval [CI]: 0.45–1.16) and 0.66 (95% CI: 0.36–1.20), respectively. There was a 27% improvement in time to progression in patients receiving the vaccine. The results were not statistically significant. However, they were significant enough to conduct a phase III trial.[35]

The MAGRIT (MAGE-A3 as Adjuvant Non–Small Cell Lung Cancer Immunotherapy) phase III trial included 2312 eligible patients with MAGE-A3–positive NSCLC. Patients were randomized to either the vaccine or placebo arm. Median DFS was 60.5 months for the MAGE-A3 arm and 57.9 months for the placebo arm. However, it was not statistically significant (hazard ratio [HR] 1.02, 95% CI: 0.89–1.18; $P = 0.74$).[36]

L-BLP25 (Stimuvax) is a peptide-based vaccine that targets Mucin 1 (MUC1). The vaccine benefits from a liposome as a vector and the adjuvant monophosphoryl lipid A, and three lipids.[37] A phase IIb trial evaluated L-BLP25 in advanced stages of NSCLC. The OS for patients in the L-BLP25 plus best supportive care (BSC) arm was 4.4 months longer than what patients in the arm receiving BSC showed ($P = 0.112$).[38] In the updated analysis released from the study, OS was 17.2 months in L-BLP25 plus BSC arm whereas it was 13.0 months in the BSC arm (HR 0.745, 95% CI: 0.533–1.042). The 3-year survival rate was 31% for the intervention arm and 17% for the BSC arm ($P = 0.035$). Moreover, in a subgroup of patients with stage IIIB locoregional tumor, a great improvement in the OS was reported in the vaccine arm compared with the BSC arm (30.6 versus 13.3 months; HR 0.548, 95% CI: 0.301–0.999).[39]

L-BLP25 compared with placebo was evaluated in a phase III trial (START trial). In this study, 1513 patients with unresectable stage III NSCLC who did not progress after primary chemotherapy and radiation were included. The median OS was 25.6 months for the L-BLP25 arm and 22.3 months for the placebo arm. However, it was not statistically significant (adjusted HR 0.88; 95% CI: 0.75–1.03; $P = 0.123$). Analysis of patients who received L-BLP25 and concurrent chemotherapy and radiation ($n = 538$) showed a significant superiority of the vaccine over placebo ($n = 268$) to increase the OS. The OS was 30.8 months for the L-BLP25 arm and 20.6 months for the placebo arm (HR 0.78; 95% CI: 0.64–0.95; $P = 0.016$).[40] Therefore, the START2 trial began, to compare the OS between patients who received either L-BLP25 or controls with concurrent chemoradiation therapy.[41] INSPIRE is also another phase III trial; the aim was to evaluate the survival benefits of L-BLP25 compared with placebo, similar to START, but in the Asian population.[42]

Belagenpumatucel-L is a tumor cell vaccine prepared from five allogeneic NSCLC cell lines, which are genetically engineered to express sequences to inhibit transforming growth factor β2 (TGF-β2). TGF-β2 is an immunosuppressive agent that has inhibitory effects on natural killer cells, activated killer cells, and DCs.[43]

Belagenpumatucel-L was evaluated at different doses in a randomized phase II trial on patients with NSCLC. Three doses (1.25×10^7, 2.5×10^7, and 5×10^7 cells/injection) were administered to subgroups. The 1- and 2-year survival rates were 39% and 20%, respectively, for patients at the low dose (1.25×10^7 cells/injection). The outcomes were 68% and 52%, respectively, for patients at the higher doses (2.5×10^7 to 5×10^7 cells/injection). The OS was 252 days at the low dose and 581 days at higher doses.[44]

After this study, a phase III trial of belagenpumatucel-L, the STOP trial,[45] was initiated to evaluate the OS in advanced stages of NSCLC (T3N2-IIIA, IIIB, and IV). In this trial, 532 patients who were previously treated with chemotherapy and had not progressed were randomized to the vaccine or placebo arms. The OS was 20.3 months for the belagenpumatucel-L arm and 17.8 months for the placebo arm (HR 0.94; $P = 0.594$). The STOP trial was not successful to achieve its primary end point. However, there were promising outcomes in the subgroups. Improvement in OS was reported for the vaccine arm compared with the placebo arm (20.7 versus 13.4 months, respectively) (HR 0.75; $P = 0.083$) in patients who were treated within 12 weeks of completing chemotherapy. Interestingly, patients with stage IIIB/IV non-adenocarcinoma randomized within 12 weeks of completing chemotherapy were reported to have improved OS in the vaccine arm versus the placebo arm (19.9 versus 12.3 months; respectively; HR 0.55; $P = 0.036$). It was more interesting in patients who were treated with radiation before the vaccine: The OS was 40.1 months for belagenpumatucel-L and 10.3 months for placebo in this subgroup (HR 0.45; $P = 0.014$). The results of subgroup analysis were promising for undertaking further studies to evaluate belagenpumatucel-L in NSCLC.

GVAX is composed of irradiated autologous or allogeneic whole tumor cells that are genetically engineered to produce recombinant GM-CSF. This vaccine was assessed in several types of cancer including prostate and NSCLC. Although, it did not show efficacy in prostate cancer, considerable clinical outcomes were demonstrated in some patients with NSCLC.[41]

In a phase I trial evaluating GVAX in NSCLC, 5 of 34 patients had stable disease, 1 had a mixed response, and 2 who had undergone prior surgical resection had a DFS of over 42 months.[46] A phase I/II trial of GVAX was carried out in 63 patients with advanced-stage disease and 20 with early stages of NSCLC. Three patients with advanced-stage disease experienced durable and complete regression for more than 5 years. Nevertheless, a phase III trial of GVAX in NSCLC was not started owing to pessimism caused by the trial of prostate cancer.[47]

VACCINES FOR COLORECTAL CANCER

Various types of cancer vaccines have been evaluated to treat colorectal cancer. Administration of autologous tumor cells with the adjuvant *Bacillus* Calmette–Guérin (BCG) is one type explained here. Patients were randomized to receive either surgical removal plus the vaccine or surgical removal alone. No statistically significant improvement were seen in the DFS or OS of the vaccine group after a median follow-up of more than 7 years.[48] However, long-term follow-up analysis resulted in statistically significant improvements in all primary end points (including OS and DFS) in patients with stage II colon cancer but not in stage III.[49]

To enhance the immunogenicity of autologous vaccines for colorectal cancer, a low pathogenic strain of the Newcastle disease virus (NDV), which is a nonlytic virus, were added to the vaccine formulation. The autologous tumor cell vaccine infected with NDV was evaluated in phase II and III trials in metastatic colorectal cancer. In the phase III trial, patients with colon or rectal cancer who had confirmed metastasis to the liver were randomized to either the vaccine or control group.[50] No improvement in OS or metastasis-free survival was reported in the vaccine group. Nevertheless, there was a significant improvement in both OS and metastasis-free survival in patients with colon cancer compared with rectal cancer in the intention-to-treat analysis. Altogether, administration of autologous vaccines in colon cancer is supported by some evidence, but it is not powerful for rectal cancer. Patients with stage II colorectal cancer are better candidates for vaccine than those with stage III.

Peptide vaccines were also evaluated for colorectal cancer. In a phase II trial in patients with metastatic colorectal cancer, a five-peptide cocktail vaccine was evaluated. It was safe but not capable of inducing clinical responses.[51] A seven-peptide cocktail vaccine was evaluated in colorectal cancer as well, which elicited positive clinical responses. A longer OS was also reported in those patients, who developed cytotoxic T-lymphocyte responses to all seven peptides.[52] Further clinical trials are being conducted to evaluate poly-epitope peptide vaccines plus adjuvants.[53]

VACCINES FOR BREAST CANCER

Cancer vaccines have been evaluated for breast cancer and have shown promising results among immunotherapeutics. Peptide-based and genetic-based cancer vaccines demonstrated interesting outcomes in early phases, which resulted in further clinical trials being conducted.[54]

NeuVax (Nelipepimut-S or E75) is a peptide vaccine targeting human epidermal growth factor receptor 2 (HER2), which is a central tumor antigen in breast cancer. HER2 is responsible for activating a number of signaling pathways that have a role in proliferation, angiogenesis, and invasion in breast cancer.[55] It also correlates with negative clinical responses.[56] The vaccine has been designed against the peptide E75 and is combined with GM-CSF.[57]

NeuVax was shown to be safe and to induce an immune response in a phase I trial.[58] It was tested in a phase II trial, which showed an improvement in OS and DFS in patients with human leukocyte antigen (HLA)-A2+ or HLA-A3+.[57,59] Now it is the only cancer vaccine for breast cancer that is being assessed in a phase III trial. The PRESENT trial (Prevention of Recurrence in Early-Stage,

Node-Positive Breast Cancer With Low to Intermediate HER2 Expression With NeuVax Treatment) included 758 patients to assess the efficacy of NeuVax plus GM-CSF in improving OS and DFS compared with standard of care plus GM-CSF. The trial was completed but the results have yet to be released (NCT01479244). The vaccine is being evaluated in a number of phase II trials, such as combined with trastuzumab, an approved monoclonal antibody against HER2 (NCT01570036).

REFERENCES

1. Schlom J. Therapeutic cancer vaccines: current status and moving forward. *J Natl Cancer Inst* 2012;**104**(8): 599–613.
2. von Mehren M, et al. Pilot study of a dual gene recombinant avipox vaccine containing both carcinoembryonic antigen (CEA) and B7.1 transgenes in patients with recurrent CEA-expressing adenocarcinomas. *Clin Cancer Res* 2000;**6**(6):2219–28.
3. von Mehren M, et al. The influence of granulocyte macrophage colony-stimulating factor and prior chemotherapy on the immunological response to a vaccine (ALVAC-CEA B7.1) in patients with metastatic carcinoma. *Clin Cancer Res* 2001;**7**(5):1181–91.
4. Hsieh C-L, Chen D-S, Hwang L-H. Tumor-induced immunosuppression: a barrier to immunotherapy of large tumors by cytokine-secreting tumor vaccine. *Hum Gene Ther* 2000;**11**(5):681–92.
5. Mendez R, et al. Identification of different tumor escape mechanisms in several metastases from a melanoma patient undergoing immunotherapy. *Cancer Immunol Immunother* 2007;**56**(1):88–94.
6. Berger M, et al. Phase I study with an autologous tumor cell vaccine for locally advanced or metastatic prostate cancer. *J Pharm Pharmaceut Sci* 2007;**10**(2):144–52.
7. Ruttinger D, et al. Adjuvant therapeutic vaccination in patients with non-small cell lung cancer made lymphopenic and reconstituted with autologous PBMC: first clinical experience and evidence of an immune response. *J Transl Med* 2007;**5**:43.
8. de Weger VA, et al. Clinical effects of adjuvant active specific immunotherapy differ between patients with microsatellite-stable and microsatellite-instable colon cancer. *Clin Cancer Res* 2012;**18**(3):882–9.
9. Schneble EJ, et al. The HER2 peptide nelipepimut-S (E75) vaccine (NeuVax) in breast cancer patients at risk for recurrence: correlation of immunologic data with clinical response. *Immunotherapy* 2014;**6**(5):519–31.
10. Plate J. Clinical trials of vaccines for immunotherapy in pancreatic cancer. *Expert Rev Vaccines* 2011;**10**(6): 825–36.
11. Leffers N, et al. Vaccine-based clinical trials in ovarian cancer. *Expert Rev Vaccines* 2011;**10**(6):775–84.
12. Pyzer AR, Avigan DE, Rosenblatt J. Clinical trials of dendritic cell-based cancer vaccines in hematologic malignancies. *Hum Vaccines Immunother* 2014;**10**(11):3125–31.
13. Alexandrov LB, et al. Signatures of mutational processes in human cancer. *Nature* 2013;**500**(7463):415–21.
14. Halabi S, et al. Prognostic model for predicting survival in men with hormone-refractory metastatic prostate cancer. *J Clin Oncol* 2003;**21**(7):1232–7.
15. Patel PH, Kockler DR. Sipuleucel-T: a vaccine for metastatic, asymptomatic, androgen-independent prostate cancer. *Ann Pharmacother* 2008;**42**(1):91–8.
16. Di Lorenzo G, Ferro M, Buonerba C. Sipuleucel-T (Provenge(R)) for castration-resistant prostate cancer. *BJU Int* 2012;**110**(2 Pt 2):E99–104.
17. Schepisi G, et al. Immunotherapy for prostate cancer: where we are headed. *Int J Mol Sci* 2017;**18**(12):2627.
18. Kantoff PW, et al. Sipuleucel-T immunotherapy for castration-resistant prostate cancer. *N Engl J Med* 2010; **363**(5):411–22.

19. Schellhammer PF, et al. Lower baseline prostate-specific antigen is associated with a greater overall survival benefit from sipuleucel-T in the Immunotherapy for Prostate Adenocarcinoma Treatment (IMPACT) trial. *Urology* 2013;**81**(6):1297–302.
20. Madan RA, et al. Prostvac-VF: a vector-based vaccine targeting PSA in prostate cancer. *Expet Opin Invest Drugs* 2009;**18**(7):1001–11.
21. Eder JP, et al. A phase I trial of a recombinant vaccinia virus expressing prostate-specific antigen in advanced prostate cancer. *Clin Cancer Res* 2000;**6**(5):1632–8.
22. Kantoff PW, et al. Overall survival analysis of a phase II randomized controlled trial of a Poxviral-based PSA-targeted immunotherapy in metastatic castration-resistant prostate cancer. *J Clin Oncol* 2010;**28**(7): 1099–105.
23. Rojas-Martinez A, et al. Intraprostatic distribution and long-term follow-up after AdV-tk immunotherapy as neoadjuvant to surgery in patients with prostate cancer. *Cancer Gene Ther* 2013;**20**(11):642–9.
24. Rosenberg SA, Yang JC, Restifo NP. Cancer immunotherapy: moving beyond current vaccines. *Nat Med* 2004;**10**(9):909–15.
25. Schwartzentruber DJ, et al. gp100 peptide vaccine and interleukin-2 in patients with advanced melanoma. *N Engl J Med* 2011;**364**(22):2119–27.
26. McDermott D, et al. Efficacy and safety of ipilimumab in metastatic melanoma patients surviving more than 2 years following treatment in a phase III trial (MDX010-20). *Ann Oncol* 2013;**24**(10):2694–8.
27. Testori A, et al. Phase III comparison of vitespen, an autologous tumor-derived heat shock protein gp96 peptide complex vaccine, with physician's choice of treatment for stage IV melanoma: the C-100-21 Study Group. *J Clin Oncol* 2008;**26**(6):955–62.
28. Schadendorf D, et al. Dacarbazine (DTIC) versus vaccination with autologous peptide-pulsed dendritic cells (DC) in first-line treatment of patients with metastatic melanoma: a randomized phase III trial of the DC study group of the DeCOG. *Ann Oncol* 2006;**17**(4):563–70.
29. Linette GP, Carreno BM. Neoantigen vaccines pass the immunogenicity test. *Trends Mol Med* 2017;**23**(10): 869–71.
30. Lu YC, Robbins PF. Targeting neoantigens for cancer immunotherapy. *Int Immunol* 2016;**28**(7):365–70.
31. Ott PA, et al. An immunogenic personal neoantigen vaccine for patients with melanoma. *Nature* 2017; **547**(7662):217–21.
32. Sahin U, et al. Personalized RNA mutanome vaccines mobilize poly-specific therapeutic immunity against cancer. *Nature* 2017;**547**(7662):222–6.
33. Sienel W, et al. Melanoma associated antigen (MAGE)-A3 expression in Stages I and II non-small cell lung cancer: results of a multi-center study. *Eur J Cardio Thorac Surg* 2004;**25**(1):131–4.
34. Shigematsu Y, et al. Clinical significance of cancer/testis antigens expression in patients with non-small cell lung cancer. *Lung Cancer* 2010;**68**(1):105–10.
35. Vansteenkiste J, et al. Final results of a multi-center, double-blind, randomized, placebo-controlled phase II study to assess the efficacy of MAGE-A3 immunotherapeutic as adjuvant therapy in stage IB/II non-small cell lung cancer (NSCLC). *J Clin Oncol* 2007;**25**(18_Suppl.). 7554–7554.
36. Vansteenkiste JF, et al. Efficacy of the MAGE-A3 cancer immunotherapeutic as adjuvant therapy in patients with resected MAGE-A3-positive non-small-cell lung cancer (MAGRIT): a randomised, double-blind, placebo-controlled, phase 3 trial. *Lancet Oncol* 2016;**17**(6):822–35.
37. Sangha R, Butts C. L-BLP25: a peptide vaccine strategy in non small cell lung cancer. *Clin Cancer Res* 2007; **13**(15 Pt 2). s4652-4.
38. Butts C, et al. Randomized phase IIB trial of BLP25 liposome vaccine in stage IIIB and IV non-small-cell lung cancer. *J Clin Oncol* 2005;**23**(27):6674–81.

39. Butts C, et al. Updated survival analysis in patients with stage IIIB or IV non-small-cell lung cancer receiving BLP25 liposome vaccine (L-BLP25): phase IIB randomized, multicenter, open-label trial. *J Cancer Res Clin Oncol* 2011;**137**(9):1337–42.
40. Butts CA, et al. START: a phase III study of L-BLP25 cancer immunotherapy for unresectable stage III non-small cell lung cancer. *J Clin Oncol* 2013;**31**(15_Suppl.). 7500–7500.
41. Ramlogan-Steel CA, Steel JC, Morris JC. Lung cancer vaccines: current status and future prospects. *Transl Lung Cancer Res* 2014;**3**(1):46–52.
42. Wu YL, et al. INSPIRE: a phase III study of the BLP25 liposome vaccine (L-BLP25) in Asian patients with unresectable stage III non-small cell lung cancer. *BMC Cancer* 2011;**11**:430.
43. Nemunaitis J, Murray N. Immune-modulating vaccines in non-small cell lung cancer. *J Thorac Oncol* 2006; **1**(7):756–61.
44. Nemunaitis J, et al. Phase II study of belagenpumatucel-L, a transforming growth factor beta-2 antisense gene-modified allogeneic tumor cell vaccine in non-small-cell lung cancer. *J Clin Oncol* 2006;**24**(29): 4721–30.
45. Giaccone G, Bazhenova L, Nemunaitis J. *A phase III study of belagenpumatucel-L therapeutic tumor cell vaccine for non-small cell lung cancer* [abstract LBA 2]. 2013 European Cancer Congress; Presented September 28, 2013.
46. Salgia R, et al. Vaccination with irradiated autologous tumor cells engineered to secrete granulocyte-macrophage colony-stimulating factor augments antitumor immunity in some patients with metastatic non-small-cell lung carcinoma. *J Clin Oncol* 2003;**21**(4):624–30.
47. Nemunaitis J, et al. Granulocyte-macrophage colony-stimulating factor gene-modified autologous tumor vaccines in non-small-cell lung cancer. *J Natl Cancer Inst* 2004;**96**(4):326–31.
48. Vermorken JB, et al. Active specific immunotherapy for stage II and stage III human colon cancer: a randomised trial. *Lancet* 1999;**353**(9150):345–50.
49. Uyl-de Groot CA, et al. Immunotherapy with autologous tumor cell-BCG vaccine in patients with colon cancer: a prospective study of medical and economic benefits. *Vaccine* 2005;**23**(17–18):2379–87.
50. Schulze T, et al. Efficiency of adjuvant active specific immunization with Newcastle disease virus modified tumor cells in colorectal cancer patients following resection of liver metastases: results of a prospective randomized trial. *Cancer Immunol Immunother* 2009;**58**(1):61–9.
51. Hazama S, et al. A phase Iotal study of five peptides combination with oxaliplatin-based chemotherapy as a first-line therapy for advanced colorectal cancer (FXV study). *J Transl Med* 2014;**12**:108.
52. Okuno K, et al. Clinical trial of a 7-peptide cocktail vaccine with oral chemotherapy for patients with metastatic colorectal cancer. *Anticancer Res* 2014;**34**(6):3045–52.
53. Pol J, et al. Trial watch: peptide-based anticancer vaccines. *OncoImmunology* 2015;**4**(4):e974411.
54. Benedetti R, et al. Breast cancer vaccines: new insights. *Front Endocrinol* 2017;**8**:270.
55. Iqbal N, Iqbal N. Human epidermal growth factor receptor 2 (HER2) in cancers: overexpression and therapeutic implications. *Mol Biol Int* 2014;**2014**:9.
56. Slamon D, et al. Human breast cancer: correlation of relapse and survival with amplification of the HER-2/neu oncogene. *Science* 1987;**235**(4785):177–82.
57. Peoples GE, et al. Clinical trial results of a HER2/neu (E75) vaccine to prevent recurrence in high-risk breast cancer patients. *J Clin Oncol* 2005;**23**(30):7536–45.
58. Murray JL, et al. Toxicity, immunogenicity, and induction of E75-specific tumor-lytic CTLs by HER-2 peptide E75 (369-377) combined with granulocyte macrophage colony-stimulating factor in HLA-A2+ patients with metastatic breast and ovarian cancer. *Clin Cancer Res* 2002;**8**(11):3407–18.
59. Peoples GE, et al. Combined clinical trial results of a HER2/neu (E75) vaccine for the prevention of recurrence in high-risk breast cancer patients: U.S. Military Cancer Institute Clinical Trials Group Study I-01 and I-02. *Clin Cancer Res* 2008;**14**(3):797–803.

CHAPTER

OBSTACLES IN THE DEVELOPMENT OF THERAPEUTIC CANCER VACCINES

12

Mahsa Keshavarz-Fathi[1,2,3], **Nima Rezaei**[3,4,5]

School of Medicine, Tehran University of Medical Sciences, Tehran, Iran[1]*; Cancer Immunology Project (CIP), Universal Scientific Education and Research Network (USERN), Tehran, Iran*[2]*; Research Center for Immunodeficiencies, Children's Medical Center, Tehran University of Medical Sciences, Tehran, Iran*[3]*; Department of Immunology, School of Medicine, Tehran University of Medical Sciences, Tehran, Iran*[4]*; Network of Immunity in Infection, Malignancy and Autoimmunity (NIIMA), Universal Scientific Education and Research Network (USERN), Tehran, Iran*[5]

TUMOR ANTIGENS

The identification of an ideal target is the central issue in designing the most specific biologic therapies including immunotherapy. In the case of cancer, it might be more challenging because tumor cells originate from normal cells of body. Therefore, it is necessary to examine various targets that specifically derive an immune response against the tumor and do not harm normal cells. Tumor antigens are critical players in designing a cancer vaccine because various types of these molecules are expressed by tumor cells, which can serve as a target for vaccines. The antigens must be presented by major histocompatibility complex (MHC) molecules of patients to T cells and induce appropriate immune responses. After *in vitro* and *in vivo* validation of the immunogenicity of a tumor antigen, clinical trials are carried out to determine the safety and efficacy of the vaccine designed to target the tumor antigen.[1]

Some concerns exist regarding the identification and selection of tumor antigens. Despite infectious diseases, tumors do not contain a foreign antigen and they share most of their antigens with normal cells. A number of cancers are associated with infectious agents, such as cervical cancer associated with human papillomavirus infection and hepatocellular carcinoma associated with hepatitis B virus. In these cases, a vaccine targeting the viral agent can prevent development of the cancer. However, a few malignancies are linked to viral infection. Tumor antigens are mainly selected among the overexpressed antigens, cancer/testes antigens, which have been ignored by the immune system, and neoantigens resulted from mutations.[2,3]

Problems are caused by both self-antigens and neoantigens. Regarding the self antigens, immune tolerance might exist and the candidate antigen might be the case of down regulation during cancer

Vaccines for Cancer Immunotherapy. https://doi.org/10.1016/B978-0-12-814039-0.00012-6

development leading to escape from the immune system and failure of cancer vaccine. The neoantigens may be different among individuals. Their identification is challenging as well.

Approaches to increase the immunogenicity of self-antigens, such as their combination with novel adjuvants and development of unchallenging approaches to identify the best neoantigens serving as cancer vaccines, are recommendations that have been put forward and undertaken to achieve the goals of vaccination.[2,4]

TUMOR BURDEN

Various immunosuppressive factors in microenvironment of tumors with a high burden have a role in inhibiting the stimulation and activation of immune responses. Different cells including regulatory T cells, myeloid-derived suppressor cells (MDSCs), and tumor-associated macrophages are key inhibitory cells that function against development of a robust immune response.[5–7] Production of inhibitory cytokines and factors such as interleukin (IL)-10, transforming growth factor-β (TGF -β), and vascular endothelial growth factor (VEGF) also aggravates the immunosuppressive profile of the milieu and invasiveness of the tumor.[8,9] The phenotype of tumor cells is affected by the microenvironment. For instance, IL-8 can direct carcinoma cells toward the epithelial–mesenchymal transition (EMT) phenotype and development of stem-like properties.[10,11] A high tumor burden can induce immunosuppression by producing cytokines with a role in developing MDSCs.[12]

In preclinical studies with animals, the effect of the tumor burden on the efficacy of vaccines was reported.[13] There is also evidence from clinical trials that demonstrates differences in clinical outcomes between patients with minimal or no residual lymphoma and those with macroscopic residual lymphoma. An idiotype vaccine was administered to patients and idiotype-specific T-cell responses were assessed. In concordance with preclinical studies, 9 of 16 patients, who had no clinically evident tumors or minimal or residual tumor developed T-cell responses compared with none of the 9 patients from the subgroup with residual lymphoma.[14] This highlights the importance of tumor burden for conducting clinical trials, in which patients who were recruited had great benefit from vaccines. Therapies that are used before immunotherapy might be able to increase vaccine efficacy by reducing the tumor burden.[15]

Immunosuppressive factors and the microenvironment are considered remarkable points that determine the achievement of desired clinical outcomes in trials evaluating cancer vaccines.[16–18] Preclinical studies highlighted the effects of bulky tumors in developing such a setting. A large mass of tumoral cells can increase interstitial pressure and consequently restrict the infiltration of antibodies and effector T cells.[19,20] The lack of costimulatory molecules, which is common in solid tumors, also has a role in shaping immunosuppression. In case of the lack of these molecules, T cells, especially low-avidity T cells specific for self-antigens, develop anergy to the antigen and are unable to exert cytotoxic functions.[21,22] There are also a number of coinhibitory molecules such as programmed death 1 (PD-1), that cause immunosuppression by developing exhausted T cells.[23]

The mechanism of action for vaccines is dissimilar to other conventional therapies such as cytotoxic agents. Vaccines target the immune system rather than having a direct effect on the tumor. Enough time is usually required for developing an immune response in this case.[24,25] It can be improved by administering booster doses as well.[26] There is a dynamic interaction between the immune system and tumor cells, and destroying tumor cells increases the cross-priming of more tumor antigens to extend the immune repertoire. To develop this process, called the antigen cascade or epitope spreading, enough time is required as well.[27] Despite cytotoxic agents, vaccines as monotherapy may not result in a significant decrease in the tumor burden. However, long-term responses against tumor cells and a reduced growth rate of tumors usually occur, commonly by prime-boosting approaches.[22,28] This can improve the overall survival, which may not be accompanied by significant differences in objective responses or progression-free survival.[29,30]

Therefore, administration of vaccine therapies while a lower tumor burden rather than a high burden is present, may result in improved outcomes. It is rational that a combination of vaccines with cytotoxic agents will improve tumor regression and diminish the tumor growth rate.[30–32] It should be taken into consideration that some clinical trials with a vaccine might be terminated owing to tumor progression before enough booster doses of vaccine are administered.[22]

Moreover, there is evidence that patients with low-grade tumors benefit more from vaccine.[33] In a trial evaluating PANVAC vaccine in colorectal cancer, patients with large-volume liver metastases showed patient benefit but no tumor response (clinical benefit). The vaccine was evaluated in another trial comparing the survival of 74 patients with colorectal cancer who had undergone metastasectomy of the liver or who had lung metastases, who received PANVAC (vaccine alone or with vaccine-modified dendritic cells [DCs]) and their contemporary control group from a prospective registry, who had undergone metastasectomy.[22] There were no significant differences between the groups in the 2-year relapse-free survival (56% for the PANVAC group, 50% for the DC-PANVAC group, and 55% for the contemporary control group) However, a significant improvement in the 2-year overall survival (OS) was observed for the vaccine group over the contemporary control group (95% versus 75%). After a median follow-up of 40 months, the OS was superior in the vaccine group as well (90% versus 47%).

Patients with more indolent tumors benefit more from vaccine. In a trial in patients with metastatic hormone-refractory prostate cancer, the results of PROSTVAC (PSA-TRICOM) vaccine were compared with docetaxel in two subgroups classified based on the Halabi nomogram: (1) patients with Halabi-predicted survival (HPS) $\geq$ 18 months and (2) patients with HPS $<$ 18 months. Patients in the first subgroup, who received the vaccine, showed a remarkable OS improvement. The observed survival was $>$37.3 months for those with HPS $\geq$18 months (median HPS 20.9 months), i.e., patients with more indolent disease ($P = 0.035$). However, in the second subgroup of the vaccine arm (median HPS of 12.3 months), the OS was 14.6 months ($P = 0.63$).[34]

CLINICAL RESPONSE VERSUS IMMUNE RESPONSE

Immunologic changes after vaccine therapy are usually employed as a surrogate end point in trials[35] For vaccines against infections, monitoring of immune responses is simply conducted by measuring antibodies. The humoral responses are the indicator of vaccine efficacy. To evaluate the efficacy of cancer vaccines, T-cell responses are important; however, there are some problems with their measurement and validation.[36,37] The following methods are used to measure T-cell responses.

Counting the number of antigen-specific T cells via MHC-peptide tetramers labeled with fluorochrome is one of the methods to measure the immune response. This method has high sensitivity to find even infrequent T cells. However, the exact peptide must be applied; it is an MHC-restricted method and does not provide information about the functional status of the antigen-specific T cells.[38]

In another method, T-cell proliferation will be assessed *in vitro* in response to antigen reexposure. It is performed by T-cell hydrogen-3—thymidine incorporation evaluation; this is the old method, which is not able to determine the T-cell subsets. To quantify the proliferation of T-cell subsets, more advanced methods measure the fluorescent dye dilution.[39,40]

The level of cytokines secreted from T cells in response to *in vitro* reexposure of antigen is also assessed. Interferon-gamma is the most well-known cytokine assessed in clinical studies through the enzyme-linked immunospot assay. This method determines the proportion of T cells that produce cytokine.[41] However, the subtypes of T cells are not specified and the results are not reproducible.[42,43] The cytokines can also be measured via intracellular cytokine flow-cytometry. The advantages of this method include evaluation of both multiple cytokine production from a single cell and determination of cell subtype.[44]

Markers of T-cell activation such as the costimulatory molecule CD137 after *in vitro* reexposure of the antigen are measured and are another indicator of developing immune response.[45] Exploiting a combination of these methods can result in more reliable immune response measurement. Nevertheless, evidence of their correlation with clinical responses is necessary.[46]

Some studies have been designed to evaluate the correlation of antigen-specific immune responses and clinical responses. Delayed-type hypersensitivity response (DTH) assay is a test that correlates directly with antigen-specific T cells in the peripheral blood and the survival rate in patients with melanoma who received autologous tumor cell vaccines.[47] DTH has also been assessed to determine antigen recall or memory responses. The magnitude of T-helper 1 immune responses and antigen cascade is also correlated with the clinical outcomes.[48] Humoral responses also showed a correlation with clinical responses. A number of tumor antigen-specific immunoglobulin G antibodies in the blood correlate with increased survival. Immune responses against several antigens rather than one antigen can improve prognosis as well.[49]

To measure antitumor responses directed by immunotherapy and not just assess vaccine potency, some studies have measured CD3+ T cell infiltrates to the tumor. Such assays are "immune signatures" that can predict the patient's survival.[49,50]

Correlations between antigen-specific immune responses and clinical responses have been reported by some trials, as mentioned, but not in all studies. There are some points that should be taken into consideration to get a better overview: (1) A lot of studies assessed only immune responses in blood (such as antigen-specific T-cell or antibody response). However, they might not be a perfect indicator for immune responses in the tumor microenvironment. (2) A few clinical studies assessed different subsets of immune cells or the presence of inhibitory cells.[51] (3) Most studies assessed the amount of antigen-specific T cells, but the avidity of antigen-specific T-cell subsets has not been considered.[52] (4) Preclinical studies revealed that antigen-specific T-cell subsets to monitor may not be the same T cells that respond to the target antigen of the vaccine.[51] Epitope spreading consequently occurres in response to the vaccine. The cytotoxic T lymphocytes destroy tumor cells and lead to the release of other tumor antigens in addition to the antigen(s) of the vaccine and development of T-cell response against them.[53–55] The latter T cells demonstrated a higher avidity and magnitude in preclinical studies. They are responsible for the cure by the vaccine.[56]

PRIOR TREATMENTS

There is evidence showing that patients who received fewer prior chemotherapeutic regimens and had a longer time elapsed since their chemotherapy developed better clinical outcomes after the vaccine's administration.[22] As mentioned, to select candidate patients for vaccine therapy tumor burden is an important factor. Fewer prior regimens of chemotherapy is also significant, because multiple regimens of chemotherapy received before the vaccine can negatively alter the response to a cancer vaccine.[57] It was reported in a phase II trial testing a genetic vaccine composed of a canarypox virus vector and the gene for carcinoembryonic antigen plus CD80, the costimulatory molecule of T cell. Development of measurable immune responses were less probable in the patients, who had progression after several chemotherapeutic regimens compared with patients with fewer chemotherapeutic regimens ($P = 0.032$).[58]

DESIGNING CLINICAL TRIALS

Clinical trials are important for evaluating the safety and efficacy of a therapeutic modality. Because there are some differences between immunotherapeutics and other medications, some points should be taken into consideration in designing a clinical trials.[59] In addition to the tumor burden and prior treatments, which affect the selection of the candidate population for vaccine therapy, setting end points or surrogate end points, which appropriately compare clinical outcomes between separate arms, is necessary to conclude a correct description of success.

The conventional approach to designing clinical trials starts from a phase I for safety evaluation toward phase III for efficacy comparison, which might lead to approval of the therapy.[60] However, achievements for immune checkpoint inhibitors received approval based on the results of a phase Ib clinical trial and subgroup analysis of a phase III clinical trial.[61,62]

Setting the appropriate end point for the clinical trials of vaccines is valuable as well. In contrast to chemotherapy, progression-free survival, a tumor-based end point, is not necessarily improved in vaccine trials, whereas an increase in OS, a patient-based end point, may result.[63,64] A study compared tumor regression and tumor growth rates in patients with metastatic hormone-refractory prostate cancer among four trials using chemotherapy and one trial on vaccine therapy. The chemotherapeutic agents showed no permanent cytotoxic effect; they had an effect contemporary with their administration. The tumor growth rate increased after termination of the cytotoxic agents' activity. The kinetics of clinical response were not the same in the vaccine trial.[32]

There is some insistence regarding evaluating the efficacy of therapeutic approaches only using the criteria of objective responses such as classic Response Evaluation Criteria in Solid Tumors, which indicates tumor regression after the therapy and assesses a decrease in the size of metastatic lesions. Progressive cancer is specified as more than a 20% increase in the cumulative size of a preexisting lesion or growth of any new lesions in this way.[65] However, there is evidence of survival improvement without substantial tumor regression or improvement in time to progression with some biologic therapies such as immunotherapy and small-molecule targeted therapies.[22]

For instance, in a phase III trial evaluating ipilimumab in metastatic melanoma, which led to its approval by the US Food and Drug Administration, improvement in the survival arm without significant improvement in time to progression was observed with ipilimumab.[66]

In that study, patients with unresectable advanced stages of melanoma were randomly treated with ipilimumab alone, gp100 vaccine alone, or their combination. The OS was 10.0 months for the combination arm compared with 6.4 months for the gp100 alone arm (hazard ratio for death = 0.68; $P < .001$). The median progression-free survival was 2.76 months in the combination arm compared with 2.86 months in the ipilimumab-alone arm, and 2.76 months in the gp100-alone arm.[66]

REFERENCES

1. Tagliamonte M, et al. Antigen-specific vaccines for cancer treatment. *Hum Vaccines Immunother* 2014; **10**(11):3332–46.
2. Bowen WS, et al. Current challenges for cancer vaccine adjuvant development. *Expet Rev Vaccine* 2018; **17**(3):207–15.
3. Buonaguro L, et al. Translating tumor antigens into cancer vaccines. *Clin Vaccine Immunol* 2011;**18**(1): 23–34.
4. Butterfield LH. Cancer vaccines. *Br Med J* 2015;**350**:h988.
5. Filipazzi P, et al. Identification of a new subset of myeloid suppressor cells in peripheral blood of melanoma patients with modulation by a granulocyte-macrophage colony-stimulation factor–based antitumor vaccine. *J Clin Oncol* 2007;**25**(18):2546–53.
6. Yang L, Zhang Y. Tumor-associated macrophages: from basic research to clinical application. *J Hematol Oncol* 2017;**10**(1):58.
7. Chaudhary B, Elkord E. Regulatory T cells in the tumor microenvironment and cancer progression: role and therapeutic targeting. *Vaccines* 2016;**4**(3):28.
8. Dennis KL, et al. Current status of IL-10 and regulatory T-cells in cancer. *Curr Opin Oncol* 2013;**25**(6): 637–45.
9. Johnson BF, et al. Vascular endothelial growth factor and immunosuppression in cancer: current knowledge and potential for new therapy. *Expert Opin Biol Ther* 2007;**7**(4):449–60.
10. Fernando RI, et al. IL-8 signaling plays a critical role in the epithelial-mesenchymal transition of human carcinoma cells. *Cancer Res* 2011;**71**(15):5296–306.
11. Zhou N, et al. IL-8 induces the epithelial-mesenchymal transition of renal cell carcinoma cells through the activation of AKT signaling. *Oncol Lett* 2016;**12**(3):1915–20.
12. Escors D. Tumour immunogenicity, antigen presentation and immunological barriers in cancer immunotherapy. *New J Sci* 2014;**2014**.
13. Malmberg KJ. Effective immunotherapy against cancer: a question of overcoming immune suppression and immune escape? *Cancer Immunol Immunother* 2004;**53**(10):879–92.
14. Timmerman JM, et al. Idiotype-pulsed dendritic cell vaccination for B-cell lymphoma: clinical and immune responses in 35 patients. *Blood* 2002;**99**(5):1517–26.
15. Gulley JL, Madan RA, Schlom J. Impact of tumour volume on the potential efficacy of therapeutic vaccines. *Curr Oncol* 2011;**18**(3):e150–7.
16. Cham CM, et al. Glucose deprivation inhibits multiple key gene expression events and effector functions in CD8+ T cells. *Eur J Immunol* 2008;**38**(9):2438–50.
17. Gajewski TF, et al. Immune resistance orchestrated by the tumor microenvironment. *Immunol Rev* 2006;**213**: 131–45.
18. Gajewski TF, Chesney J, Curriel TJ. Emerging strategies in regulatory T-cell immunotherapies. *Clin Adv Hematol Oncol* 2009;**7**(1):1–10. quiz 11-2.
19. Fukumura D, et al. Tumor microvasculature and microenvironment: novel insights through intravital imaging in pre-clinical models. *Microcirculation* 2010;**17**(3):206–25.
20. Carmeliet P, Jain RK. Principles and mechanisms of vessel normalization for cancer and other angiogenic diseases. *Nat Rev Drug Discov* 2011;**10**(6):417–27.
21. Angell TE, et al. MHC class I loss is a frequent mechanism of immune escape in papillary thyroid cancer that is reversed by interferon and selumetinib treatment in vitro. *Clin Cancer Res* 2014;**20**(23):6034–44.
22. Schlom J. Therapeutic cancer vaccines: current status and moving forward. *J Natl Cancer Inst* 2012;**104**(8): 599–613.

23. Prall F, Huhns M. The PD-1 expressing immune phenotype of T cell exhaustion is prominent in the 'immunoreactive' microenvironment of colorectal carcinoma. *Histopathology* 2017;**71**(3):366–74.
24. Kudrin A. Overview of cancer vaccines: considerations for development. *Hum Vaccines Immunother* 2012; **8**(9):1335–53.
25. Gulley JL. Therapeutic vaccines: the ultimate personalized therapy? *Hum Vaccines Immunother* 2013;**9**(1): 219–21.
26. Coventry BJ, et al. Prolonged repeated vaccine immuno-chemotherapy induces long-term clinical responses and survival for advanced metastatic melanoma. *J Immunother Cancer* 2014;**2**:9.
27. Gulley JL, et al. Role of antigen spread and distinctive characteristics of immunotherapy in cancer treatment. *J Natl Cancer Inst* 2017;**109**(4):djw261.
28. Xu Z, et al. Designing therapeutic cancer vaccine trials with delayed treatment effect. *Stat Med* 2017;**36**(4): 592–605.
29. Singh BH, Gulley JL. Immunotherapy and therapeutic vaccines in prostate cancer: an update on current strategies and clinical implications. *Asian J Androl* 2014;**16**(3):364–71.
30. Madan RA, et al. Therapeutic cancer vaccines in prostate cancer: the paradox of improved survival without changes in time to progression. *Oncologist* 2010;**15**(9):969–75.
31. Ogi C, Aruga A. Clinical evaluation of therapeutic cancer vaccines. *Hum Vaccines Immunother* 2013;**9**(5): 1049–57.
32. Stein WD, et al. Tumor regression and growth rates determined in five intramural NCI prostate cancer trials: the growth rate constant as an indicator of therapeutic efficacy. *Clin Cancer Res* 2011;**17**(4):907–17.
33. Madan RA, et al. From clinical trials to clinical practice: therapeutic cancer vaccines for the treatment of prostate cancer. *Expert Rev Vaccine* 2011;**10**(6):743–53.
34. Gulley JL, et al. Immunologic and prognostic factors associated with overall survival employing a poxviral-based PSA vaccine in metastatic castrate-resistant prostate cancer. *Cancer Immunol Immunother* 2010;**59**(5): 663–74.
35. Morse MA, et al. Biomarkers and correlative endpoints for immunotherapy trials. *Am Soc Clin Oncol Educ Book* 2013;**33**:e287.
36. Shankaran V, et al. IFNγ and lymphocytes prevent primary tumour development and shape tumour immunogenicity. *Nature* 2001;**410**:1107.
37. Smyth MJ, et al. Perforin-mediated cytotoxicity is critical for surveillance of spontaneous lymphoma. *J Exp Med* 2000;**192**(5):755–60.
38. Sims S, Willberg C, Klenerman P. MHC-peptide tetramers for the analysis of antigen-specific T cells. *Expert Rev Vaccines* 2010;**9**(7):765–74.
39. Wagner U, Burkhardt E, Failing K. Evaluation of canine lymphocyte proliferation: comparison of three different colorimetric methods with the 3H-thymidine incorporation assay. *Vet Immunol Immunopathol* 1999; **70**(3–4):151–9.
40. Malyguine A, et al. *New approaches for monitoring CTL activity in clinical trials*. New York, NY: Springer New York; 2007.
41. Malyguine AM, et al. ELISPOT assay for monitoring cytotoxic T lymphocytes (CTL) activity in cancer vaccine clinical trials. *Cells* 2012;**1**(2):111–26.
42. Maecker HT, et al. Precision and linearity targets for validation of an IFNγ ELISPOT, cytokine flow cytometry, and tetramer assay using CMV peptides. *BMC Immunol* 2008;**9**(1):9.
43. Janetzki S, et al. Results and harmonization guidelines from two large-scale international Elispot proficiency panels conducted by the Cancer Vaccine Consortium (CVC/SVI). *Cancer Immunol Immunother* 2008;**57**(3): 303–15.
44. Foster B, et al. Detection of intracellular cytokines by flow cytometry. *Curr Protoc Immunol* 2007;**78**(1) [Chapter 6]: p. Unit 6.24.1-6.24.21.

45. Wolfl M, et al. Activation-induced expression of CD137 permits detection, isolation, and expansion of the full repertoire of $CD8^+$ T cells responding to antigen without requiring knowledge of epitope specificities. *Blood* 2007;**110**(1):201–10.
46. Whiteside TL, et al. Immunologic monitoring of cellular immune responses in cancer vaccine therapy. *J Biomed Biotechnol* 2011;**2011**:370374.
47. Hsueh EC, et al. Correlation of specific immune responses with survival in melanoma patients with distant metastases receiving polyvalent melanoma cell vaccine. *J Clin Oncol* 1998;**16**(9):2913–20.
48. Disis ML. Immunologic biomarkers as correlates of clinical response to cancer immunotherapy. *Cancer Immunol Immunother* 2011;**60**(3):433–42.
49. Dang Y, Disis ML. Identification of immunologic biomarkers associated with clinical response after immune-based therapy for cancer. *Ann N Y Acad Sci* 2009;**1174**:81–7.
50. Guo C, et al. Therapeutic cancer vaccines: past, present, and future. *Adv Cancer Res* 2013;**119**:421–75.
51. Schlom J, Arlen PM, Gulley JL. Cancer vaccines: moving beyond current paradigms. *Clin Canc Res* 2007; **13**(13):3776–82.
52. Rubio V, et al. Ex vivo identification, isolation and analysis of tumor-cytolytic T cells. *Nat Med* 2003;**9**(11): 1377–82.
53. Cavacini LA, et al. Evidence of determinant spreading in the antibody responses to prostate cell surface antigens in patients immunized with prostate-specific antigen. *Clin Cancer Res* 2002;**8**(2):368–73.
54. Gulley JL, et al. Combining a recombinant cancer vaccine with standard definitive radiotherapy in patients with localized prostate cancer. *Clin Cancer Res* 2005;**11**(9):3353–62.
55. Brossart P, et al. Epitope spreading occurs in cancer patients after vaccination with a single tumor antigen. *Exp Hematol* 2000;**28**(7, Suppl. 1):83.
56. Chakraborty M, et al. External beam radiation of tumors alters phenotype of tumor cells to render them susceptible to vaccine-mediated T-cell killing. *Cancer Res* 2004;**64**(12):4328–37.
57. Madan RA, et al. Therapeutic vaccines in metastatic castration-resistant prostate cancer: principles in clinical trial design. *Expert Opin Biol Ther* 2010;**10**(1):19–28.
58. von Mehren M, et al. The influence of granulocyte macrophage colony-stimulating factor and prior chemotherapy on the immunological response to a vaccine (ALVAC-CEA B7.1) in patients with metastatic carcinoma. *Clin Cancer Res* 2001;**7**(5):1181–91.
59. Simon R. Immunotherapy is different: implications for vaccine clinical trial design. *Hum Vaccines Immunother* 2017;**13**(9):2179–84.
60. Rivalland G, Scott AM, John T. Standard of care in immunotherapy trials: challenges and considerations. *Hum Vaccines Immunother* 2017;**13**(9):2164–78.
61. Robert C, et al. Anti-programmed-death-receptor-1 treatment with pembrolizumab in ipilimumab-refractory advanced melanoma: a randomised dose-comparison cohort of a phase 1 trial. *Lancet* 2014;**384**(9948): 1109–17.
62. Weber JS, et al. Nivolumab versus chemotherapy in patients with advanced melanoma who progressed after anti-CTLA-4 treatment (CheckMate 037): a randomised, controlled, open-label, phase 3 trial. *Lancet Oncol* 2015;**16**(4):375–84.
63. Fiteni F, et al. Endpoints in cancer clinical trials. *J Vis Surg* 2014;**151**(1):17–22.
64. Hoos A. Evolution of end points for cancer immunotherapy trials. *Ann Oncol* 2012;**23**(Suppl_8):viii47–52.
65. Therasse P, et al. New guidelines to evaluate the response to treatment in solid tumors. European Organization for Research and Treatment of Cancer, National Cancer Institute of the United States, National Cancer Institute of Canada. *J Natl Cancer Inst* 2000;**92**(3):205–16.
66. Hodi FS, et al. Improved survival with ipilimumab in patients with metastatic melanoma. *N Engl J Med* 2010; **363**(8):711–23.

CHAPTER

COMBINATION THERAPY: CANCER VACCINES AND OTHER THERAPEUTICS

13

Mahsa Keshavarz-Fathi[1,2,3], **Nima Rezaei**[3,4,5]

School of Medicine, Tehran University of Medical Sciences, Tehran, Iran[1]; *Cancer Immunology Project (CIP), Universal Scientific Education and Research Network (USERN), Tehran, Iran*[2]; *Research Center for Immunodeficiencies, Children's Medical Center, Tehran University of Medical Sciences, Tehran, Iran*[3]; *Department of Immunology, School of Medicine, Tehran University of Medical Sciences, Tehran, Iran*[4]; *Network of Immunity in Infection, Malignancy and Autoimmunity (NIIMA), Universal Scientific Education and Research Network (USERN), Tehran, Iran*[5]

In advanced stages of cancer, monotherapy may not be capable of inducing powerful significant outcomes because of the high tumor burden and the influence of metastasis on the vaccine's efficacy. Combination therapy emerged in the field of cancer immunotherapy similar to other therapeutic modalities. It might be used to exert an additive function agai nst cancer, such as to decrease the volume of the tumor, or to alter the immune system and overcome barriers against developing robust immune responses by the vaccine.[1] Combination therapies can improve the antitumor responses of vaccines in clinical trials in various ways such as by upregulating major histocompatibility complex (MHC) molecules, increasing the production of tumor antigens, and giving rise to the expression of Fas, tumor necrosis factor (TNF) receptor, and TNF-related ligand receptors.[2] As it was addressed in the previous chapter, the immunosuppressive phenotype of the tumor microenvironment is considered as an important factor for evaluating the efficacy of vaccines.[3,4] Inhibition of immunosuppressive cells and cytokines can be achieved by adding other therapies as well.[5] Therapeutic agents combined with vaccines use a variety of mechanisms to induce a robust immune response against tumors. They may serve as immune adjuvants or inhibitors of suppressive cells and cytokines of the tumor microenvironment.

A variety of cancer treatments have the potential to be used combined with vaccine therapy, including conventional therapies such as chemotherapy and radiation therapy (RT), targeted biologic therapies, and other immunotherapies besides vaccines. To achieve the best outcomes in combination therapy, the population that benefit from the combination should be determined and selected, and dosage and treatment schedules should be optimized. These two can be accomplished by developing predictive biomarkers and conducting preclinical and clinical studies.[6]

CHEMOTHERAPY COMBINED WITH VACCINES

Some chemotherapeutic agents alter the immunosuppression of tumors by diminishing the amount of inhibitory immune cells such as myeloid-derived suppressor cells (MDSCs) and regulatory T cells (Tregs).[7,8] A number of trials demonstrated that administering low-dose cyclophosphamide can result in a reduction in the Treg population.[9–11]

Vaccines for Cancer Immunotherapy. https://doi.org/10.1016/B978-0-12-814039-0.00013-8

For instance, in a phase II trial of a multipeptide vaccine in renal cell carcinoma, a single dose of cyclophosphamide was administered, which led to a decrease in the frequency of Tregs and improved survival in immune responders.[12] A trend toward improving survival was also reported in a phase I/II trial in ovarian cancer evaluating the combination of a multipeptide-loaded dendritic cell (DC) vaccine and cyclophosphamide.[10] There is evidence of proliferation of some subsets of T cells by the chemotherapeutic regimen of cisplatin and vinorelbine, which resulted in an increase in the ratio of T-effector cells to Tregs.[13]

A number of chemotherapeutic agents are capable of reducing the frequency of MDSCs. One example is the irinotecan plus infusional 5-fluorouracil and leucovorin regimen, used before the DC vaccine in the mouse model of colorectal cancer, which can decrease the frequency of both Tregs and MDSCs as well as induce T-helper 1 (Th1) and cytotoxic T lymphocyte (CTL) responses.[14] Docetaxel can also decrease MDSCs by directing the myeloid cells toward an M1-like polarization to exert functions against tumor cells.[15] There are other chemotherapeutics such as 5-fluorouracil and gemcitabine, which reduce the number of MDSCs and increase CD8+ effector cells.[7]

Chemotherapeutic agents may also improve the immune-mediated destruction of tumor cells generated by vaccines. Agents including oxaliplatin and anthracyclines increase the cross-priming of tumor antigens by DCs and activate T cells.[16,17] Death tumor cells after the administration of oxaliplatin release high-mobility group protein 1 (HMGB1), a Toll-like receptor 4 (TLR4) ligand, to stimulate DCs.[18] Many chemotherapeutics such as doxorubicin, cisplatin, fluorouracil, mitomycin C, and camptothecin induce the upregulation of Fas receptor on cancerous cells and promote lymphocytes to induce Fas—FasL-dependent cytotoxicity.[19] Docetaxel is one example, which can increase immune-mediated tumor cell death through various mechanisms such as causing the expression of tumor antigens, peptide—MHC complexes, and death receptors.[15,20] Docetaxel was tested in combination with vaccines such as rilimogene galvacirepvec/rilimogene glafolivec (PROSTAVAC) in prostate vaccine, which led to a longer progression-free survival compared with the historical control.[21]

In addition to decreasing the frequency of Tregs, cyclophosphamide can enhance Th1 as well as Th17 responses.[22,23] A combination of cyclophosphamide and granulocyte macrophage colony-stimulating factor (GM-CSF) secreting tumor cells was tested in patients with advanced pancreatic cancer. Patients received either the vaccine and cyclophosphamide or the vaccine alone. The combination did not have severe adverse effects. Cyclophosphamide increased survival for 2.3—4.3 months and resulted in mesothelin-specific CD8+ T cells with higher avidity as an immune response correlated with the clinical outcome.[24]

To determine the optimal doses of vaccination and chemotherapy in clinical settings, cyclophosphamide and doxorubicin plus allogenic GM-CSF—secreting tumor cells were tested in metastatic breast cancer.[25] The clinical trial was conducted subsequent to a preclinical study in neu-N mice, which demonstrated the optimal dose for cyclophosphamide and doxorubicin plus GM-CSF—secreting tumor cells targeting human epidermal growth factor receptor 2 (HER-2)/neu peptide. The optimal dose was as following: cyclophosphamide 1 day before administration of vaccine and doxorubicin 7 days after administration of vaccine.[22] The results of the clinical trial showed that low doses of cyclophosphamide (less than 200 mg/m^2) increased delayed-type hypersensitivity (DTH) and antibody responses to HER-2/neu; however, high doses decreased these measurements. Doxorubicin at all doses showed DTH to HER-2/neu. The optimal dose for doxorubicin was 35 mg/m^2, and for cyclophosphamide it was 200 mg/m^2. The chemotherapy regimen increased the availability of the vaccine in the body as well. Patients who received the vaccine alone had a decrease in GM-CSF after each vaccination; nevertheless, the GM-CSF levels continued to be at the peak level in the combination group.[26]

The results of a multicenter phase III trial suggested a significant role for the type of chemotherapy delivery in the clinical outcomes. The START* trial examining tecemotide (L-BLP25) vaccine for 1513 patients with advanced lung cancer resulted in survival advantage for a subgroup of patients, who received vaccine after concurrent chemoradiotherapy compared with patients who received placebo after concurrent chemoradiotherapy (overall survival [OS] 30.8 versus 20.6 months, hazard ratio [HR] = 0.78; 95% confidence interval: 0.64–0.95; $P = 0.016$). Similar results were not observed in all patients, who received vaccine after chemoradiotherapy (including both concurrent and sequential chemoradiotherapy). The results of the study led to a modification in the ongoing trial of tecemotide, INSPIRE, and the initiation of the START II study to evaluate the vaccine after concurrent chemoradiotherapy.[27]

A number of trials seeking the efficacy of chemotherapy and vaccine combinations for cancer treatment revealed promising results necessitating further studies to confirm the efficacy on a larger scale. Results of combination therapy with docetaxel and different types of vaccines (peptide, DC, and genetic vaccines) have been released.[28–31]

A vaccine of 20 mixed peptides (KRM-20) plus docetaxel and dexamethasone was compared with placebo plus docetaxel and dexamethasone in a phase II trial of 51 patients with chemotherapy-naive, hormone-refractory prostate cancer. The vaccine arm showed a significant decrease in the mean percent prostate-specific antigen (PSA) levels from baseline compared with the placebo arm ($P = 0.028$, multivariate analysis of variance) as well as a decrease in the frequency of MDSCs ($P = 0.007$) and improvement of CTL response ($P = 0.007$). However, the clinical outcomes were similar between the two arms.[31]

Cyclophosphamide combined with vaccines was also tested in biliary tract cancer and advanced solid tumors.[32,33] In a randomized phase II trial of patients with biliary tract cancer, who were resistant to at least one chemotherapeutics regimen, a personalized peptide vaccine was combined with low-dose cyclophosphamide. Improved progression-free survival (6.1 versus 2.9 months; HR: 0.427; $P = 0.008$) and OS (12.1 versus 5.9 months; HR: 0.376; $P = 0.004$) were observed in the combination arm compared with the vaccine alone. Interestingly, the plasma level of interleukin (IL)-6 was increased in the vaccine arm but not the combination arm, which showed the potential role of cyclophosphamide in inhibiting immunosuppression resulting from IL-6 production.[33]

RADIATION COMBINED WITH VACCINES

Radiation therapy has been known as a standard therapy with direct cytotoxicity for many types of cancer. Similar to chemotherapy, it can enhance killing of immune-mediated tumor cells.[34] Radiation therapy alone may not be capable of inducing a strong immune response but it can synergistically improve clinical outcomes when combined with immunotherapy.[35,36] Radiation therapy induces a number of immunologic alterations that may be beneficial for enhancing the effects of cancer vaccines. The alterations are present at different levels, such as changes in the molecules on the surface of tumor cells, antigen-processing machinery, and modulation of genes involved in the survival or response to immunity.[37] Tumor cells after radiation are more sensitive to killing by CTL.[13]

If subtumoricidal doses of irradiation are used for tumor cells, alterations in the phenotype of tumor cells are developed that foster antitumor immune responses. Upregulation of tumor peptides and MHC molecules, which are vital for presentation and recognition of tumor antigens, as well as expression of receptors such as Fas, which has a role in apoptosis, and intercellular adhesion molecule-1 (ICAM1), are among the immunologic effects of radiation.[38] Costimulatory molecules, e.g., OX-40L and 4-1BBL, are upregulated after radiation.[39] These molecules diminish the frequency of Tregs and consequently their inhibitory functions. In addition to decreasing the tumor size, radiation used locally, can induce an inflammatory microenvironment that stimulates the presentation of tumor antigens by DCs.[40]

Radiation therapy (RT) can be used as an immunomodulatory modality, as supported by some evidence. In a number of *in vivo* models, radiation led to the restriction of immunosuppressive cells and a subsequent improvement in the vaccine's efficacy.[41]

Combined RT and cancer vaccines were evaluated in a lot of preclinical and clinical studies., *and several trials were conducted.* A phase I study is being carried out to evaluate the combination of three different therapies for pancreatic adenocarcinoma after surgical removal. Immunotherapy with GVAX vaccine, RT with fractionated stereotactic body RT (SBRT) 6.6 Gy, and chemotherapy with FOLFIRINOX regimen with or without low-dose cyclophosphamide were administered to 19 patients. SBRT is being administered in two arms over 5 days starting within 6–10 weeks after surgery (arm 1) or 13–17 days after the first vaccination (arm 2). The primary end point is the safety of the combination and the secondary end points include OS, disease-free survival, and distant metastases-free survival (NCT01595321). However, it is hard to assess the effects of radiation when using both chemotherapy and RT.

Combination of conformal RT and DC vaccine loaded with tumor-cell lysates or peptides was examined in patients with various types of cancer including metastatic, recurrent, or locally advanced cancers of head and neck, esophagus, lung, pancreas, and uterus. In the trial, 40 patients received intensity-modulated RT through tomotherapy, SBRT, or three-dimensional conformal RT. For patients who were previously treated with RT, a total dose of 30 Gy, and for others a total dose of 60 Gy (at standard 2 Gy/fraction) were administered. Afterward, DC vaccines were used every other week, up to seven times. A response rate of 61% was observed in 31 patients who were treated with full-dose RT. A response rate of 55% was reported at 30 Gy RT. In nine patients evaluable for tumor response, 22% had partial response, 33% had stable disease, and 44% had progressive disease, based on the response evaluation criteria in solid tumors.[42]

External-beam RT (EBRT) was tested in combination with PROSTAVAC in patients with localized prostate cancer. In this phase I trial, improvements in PSA-specific T-cell responses were reported for patients in the combination group compared with EBRT alone.[43] The priming vaccination consisted of recombinant vaccinia (rV) with the gene encoding PSA and B7.1. The boosting vaccination, administered seven times each 28-day cycle, was recombinant fowlpox (rF) with the gene encoding PSA.

On day 2 of each 28-day cycle, vaccines were administered and subcutaneous GM-CSF was used at the vaccination site on days 1–4 and subcutaneous IL-2 was used at the abdomen on days 8–12. Between the fourth and sixth vaccinations, standard EBRT (i.e., $\geq$70 Gy, with 2.0 Gy/fraction) was used. An increase in PSA-specific T cells was reported in 13 of 17 patients in the combination group but not in the RT group.[44] However, a follow-up study showed no superiority for the combination over the standard therapy. Among 12 patients, there was one, who had a PSA-specific immune response lasting 66 months after enrollment.[45]

To enhance the efficacy of poxviral vaccines, two other costimulatory molecules were added and a triad of costimulatory molecules (TRICOM) with more efficacy was generated.[46] The vaccine rV/F-CEA/TRICOM combined with RT was examined in a trial of 12 patients with advanced gastrointestinal cancer, who had metastasis to the liver. On day 1, patients received the vaccine with recombinant vaccinia, i.e., rV-CEA/TRICOM and the boosters were administered with the recombinant fowlpox, rF-CEA/TRICOM, biweekly. A total dose of 32 Gy radiation, starting on day 21 and applied in the split courses, was administered. Adjuvant was administered as rF–GM-CSF for all vaccines. The combination was safe even in patients, who had bulky tumor and received many chemotherapeutics before the study. Two patients showed stable disease for 5 months.[47] Induction of an antigen cascade was also reported in a trial combining RT and PROSTAVAC. Although the vaccine targeted PSA, specific immune responses against prostatic acid phosphatase (PAP), prostate stem cell antigen (PSCA), prostate-specific membrane antigen (PSMA), and MUC1 were developed.[43]

Several trials were conducted to assess the safety and efficacy of vaccine and RT combinations such as the combination of the approved vaccine sipuleucel-T and different radiation strategies (NCT01807065, NCT01818986, and NCT01833208).

TARGETED THERAPIES COMBINED WITH IMMUNOTHERAPY

The efficacy of cancer vaccines might be improved when they are combined with targeted therapies such as proteasome inhibitors, mechanistic target of rapamycin inhibitors, or some tyrosine kinase inhibitors (TKIs), which influence antitumor immune responses.[6] Similar to chemotherapy, small molecule inhibitors can decrease the frequency of immunosuppressive cells (Tregs and MDSCs) and improve the function of DCs and T cells.[48] The multipeptide vaccine IMA901 has been tested combined with sunitinib and a single dose of cyclophosphamide. The combination entered a phase III trial as a first-line treatment for metastatic renal cell carcinoma (NCT01265901).

TKIs are also potential targets for combination with cancer vaccines because they can inhibit the signal transducer and activator of transcription (STAT) signaling pathways and restrict immunosuppression through alterations in Tregs, MDSCs, and coinhibitory molecules.[49–52]

HORMONE THERAPY COMBINED WITH VACCINE

In the case of hormone-sensitive prostate cancer, usually androgen deprivation therapy (ADT) is undertaken. There is evidence of alterations in the immune system caused by this therapy.[53,54] ADT can improve thymic regeneration and the frequency of prostate infiltrating immune cells such as T cells (CD4+ and CD8+), natural killer cells, and macrophages.[55,56] The combination of ADT, PROSTAVAC, and docetaxel compared with ADT alone will be tested in patients with metastatic castrate-sensitive prostate cancer as a phase II trial (NCT02649855).

VACCINE COMBINED WITH OTHER IMMUNOTHERAPEUTIC MODALITIES

A lot of immunotherapeutics, including active and passive approaches, were approved for cancer. The approved modalities, reviewed in Chapter 1, use various mechanisms of action to induce or empower antitumor immune responses. A combination of these modalities might lead to superior effects over monotherapy as a result of using several modes of action. Cancer vaccines have been combined with immune checkpoint inhibitors, monoclonal antibodies, immune stimulants or adjuvants, and immunomodulatory strategies.[6,57,58]

Cytokines such as IL-15, IL-2, GM-CSF, interferons, and TLR[59–63] agonists are among the adjuvants evaluated for use alone or combined with vaccines for cancer treatment. Ligands of other receptors of innate immune system such as nucleotide oligomerization domain-like receptors and retinoic acid-inducible gene-like receptors (reviewed in Chapter 3) can serve as an adjuvant as well.[64,65] The new adjuvants should be tested to demonstrate their superiority over the conventional types in augmenting immune responses.

The clinical efficacy of immune checkpoint inhibitors such as anti-CTLA4 antibodies or anti-PD1/PD-L1 monoclonal antibodies (mAb) is improved by enhancing tumor-infiltrating lymphocytes.[66] Moreover, to overcome the immunosuppressive setting of the tumor microenvironment and improve the efficacy of vaccines, a blockade of coinhibitory molecules is beneficial. Combining these two approaches resulted in additive and/or synergistic effects in preclinical studies in different cancers such as aggressive neuroblastoma and breast tumor models.[67,68] However, the clinical trials did not show positive outcomes in the combination arms. For example, in a trial examining the combination of long peptide vaccine and ipilimumab or vemurafenib in unresectable stage III or IV melanoma, a clinical response did not occur despite the induction of T-cell responses.[69]

A known phase III trial in patients with unresectable advanced melanoma demonstrated the effects of combining vaccine and anti-CTLA-4 mAb. In the study, patients were randomized to one of three arms: i.e., ipilimumab plus placebo, gp100 vaccine plus placebo, or the combination of gp100 and ipilimumab. The OS for the combination arm was 10 months, compared with 6.4 months for the gp100 arm (HR for death, 0.68; $P < 0.001$). The overall survival was 10.1 months for the ipilimumab arm (HR for death compared with gp100 alone, 0.66; $P = 0.003$).[70]

Clinical trials on neoantigen vaccines showed promising results, which was a glimmer of success in the field of cancer vaccine. Combined checkpoint inhibitors and neoantigen vaccines are also being evaluated in early clinical trials. For instance, NeoVax, a personalized neoantigen cancer vaccine combined with ipilimumab, an anti-CTLA-4 monoclonal antibody, will be administered to patients with kidney cancer (NCT02950766).

Many attempts to block immune checkpoint molecules have been made to target CTLA4 and PD-1 molecules, whereas new checkpoint inhibitors are being developed to target other immune checkpoints such as LAG-3 and Tim-3. Inhibition of LAG-3 is being tested in a phase I/IIa clinical trial administering a cancer vaccine containing anti-LAG-3 antibody and five synthetic peptides derived from melanoma tumor antigens to target both CD8+ and CD4+ T cells (NCT01308294).

In addition to immune checkpoints, some immunomodulatory molecules can affect antitumor immune responses. Agonistic monoclonal antibodies toward immunomodulatory molecules such as 4-1BB, OX40, and CD40 can be combined with cancer vaccines. The results of targeting these molecules were positive in preclinical and clinical studies.[71–73] For instance, in mouse tumor models, adding a CD40 agonist to the synthetic peptide vaccine and TLR agonists resulted in antigen-specific CD8+ T cells and improved protective and therapeutic effects.[71]

mAbs are also used to target special molecules on the tumor cells or even immune cells. For instance, the mAb against CD25 targets Tregs to hamper immunosuppression and may potentially be used in combination with cancer vaccines. However, it did not improve the vaccine's efficacy in a clinical trial.[74,75]

A lot of mAbs are approved for the treatment of various types of cancer (reviewed in Chapter 1). They have been found to have potential for combination therapy with vaccines. Many phase II clinical trials are examining the efficacy of trastuzumab, a mAb targeting the HER-2/neu molecule, in combination with vaccines for patients with HER-2/neu+ metastatic breast cancer. In one trial, trastuzumab combined with allogeneic GM-CSF–secreting tumor cells and cyclophosphamide is being tested in 20 patients to assess safety and the immunologic response (NCT00847171). The next trial is a combination of trastuzumab and peptide vaccine (NCT00343109). The third example combined trastuzumab with a DC vaccine and vinorelbine (NCT00266110).

In this international study, the investigators faced many challenges, including a study suspension, which complicated the conduct of the study and made the results difficult to interpret. https://www.ncbi.nlm.nih.gov/pmc/articles/PMC4073360/.

REFERENCES

1. Bilusic M, Madan RA. Therapeutic cancer vaccines: the latest advancement in targeted therapy. *Am J Therapeut* 2012;**19**(6):e172–e81.
2. Gulley JL, Madan RA, Arlen PM. Enhancing efficacy of therapeutic vaccinations by combination with other modalities. *Vaccine* 2007;**25**(Suppl. 2):B89–96.
3. Takahashi T, Sakaguchi S. The role of regulatory T cells in controlling immunologic self-tolerance. *Int Rev Cytol* 2003;**225**:1–32.
4. Mule JJ, et al. Transforming growth factor-beta inhibits the in vitro generation of lymphokine-activated killer cells and cytotoxic T cells. *Cancer Immunol Immunother* 1988;**26**(2):95–100.
5. Hodge JW, et al. The tipping point for combination therapy: cancer vaccines with radiation, chemotherapy, or targeted small molecule inhibitors. *Semin Oncol* 2012;**39**(3):323–39.
6. Apetoh L, et al. Combining immunotherapy and anticancer agents: the right path to achieve cancer cure? *Ann Oncol* 2015;**26**(9):1813–23.
7. Vincent J, et al. 5-Fluorouracil selectively kills tumor-associated myeloid-derived suppressor cells resulting in enhanced T cell-dependent antitumor immunity. *Cancer Res* 2010;**70**(8):3052–61.
8. Li J-Y, et al. Selective depletion of regulatory T cell subsets by docetaxel treatment in patients with nonsmall cell lung cancer. *J Immunol Res* 2014;**2014**:10.
9. Powell E, Chow LQ. BLP-25 liposomal vaccine: a promising potential therapy in non-small-cell lung cancer. *Expert Rev Respir Med* 2008;**2**(1):37–45.
10. Chu CS, et al. Phase I/II randomized trial of dendritic cell vaccination with or without cyclophosphamide for consolidation therapy of advanced ovarian cancer in first or second remission. *Cancer Immunol Immunother* 2012;**61**(5):629–41.
11. Le DT, Jaffee EM. Regulatory T-cell modulation using cyclophosphamide in vaccine approaches: a current perspective. *Cancer Res* 2012;**72**(14):3439–44.
12. Walter S, et al. Multipeptide immune response to cancer vaccine IMA901 after single-dose cyclophosphamide associates with longer patient survival. *Nat Med* 2012;**18**(8):1254–61.
13. Gameiro SR, et al. Exploitation of differential homeostatic proliferation of T-cell subsets following chemotherapy to enhance the efficacy of vaccine-mediated antitumor responses. *Cancer Immunol Immunother* 2011;**60**(9):1227–42.
14. Kim HS, et al. Dendritic cell vaccine in addition to FOLFIRI regimen improve antitumor effects through the inhibition of immunosuppressive cells in murine colorectal cancer model. *Vaccine* 2010;**28**(49):7787–96.
15. Kodumudi KN, et al. A novel chemoimmunomodulating property of docetaxel: suppression of myeloid-derived suppressor cells in tumor bearers. *Clin Cancer Res* 2010;**16**(18):4583–94.

16. Tesniere A, et al. Immunogenic death of colon cancer cells treated with oxaliplatin. *Oncogene* 2010;**29**(4): 482–91.
17. Kepp O, et al. Molecular determinants of immunogenic cell death elicited by anticancer chemotherapy. *Cancer Metastasis Rev* 2011;**30**(1):61–9.
18. Apetoh L, et al. Toll-like receptor 4-dependent contribution of the immune system to anticancer chemotherapy and radiotherapy. *Nat Med* 2007;**13**(9):1050–9.
19. Micheau O, et al. Sensitization of cancer cells treated with cytotoxic drugs to fas-mediated cytotoxicity. *J Natl Cancer Inst* 1997;**89**(11):783–9.
20. Garnett CT, Schlom J, Hodge JW. Combination of docetaxel and recombinant vaccine enhances T-cell responses and antitumor activity: effects of docetaxel on immune enhancement. *Clin Cancer Res* 2008;**14**(11): 3536–44.
21. Arlen PM, et al. A randomized phase II study of concurrent docetaxel plus vaccine versus vaccine alone in metastatic androgen-independent prostate cancer. *Clin Cancer Res* 2006;**12**(4):1260–9.
22. Machiels JP, et al. Cyclophosphamide, doxorubicin, and paclitaxel enhance the antitumor immune response of granulocyte/macrophage-colony stimulating factor-secreting whole-cell vaccines in HER-2/neu tolerized mice. *Cancer Res* 2001;**61**(9):3689–97.
23. Viaud S, et al. Cyclophosphamide induces differentiation of Th17 cells in cancer patients. *Cancer Res* 2011; **71**(3):661–5.
24. Laheru D, et al. Allogeneic granulocyte macrophage colony-stimulating factor-secreting tumor immunotherapy alone or in sequence with cyclophosphamide for metastatic pancreatic cancer: a pilot study of safety, feasibility, and immune activation. *Clin Cancer Res* 2008;**14**(5):1455–63.
25. Emens LA, et al. Timed sequential treatment with cyclophosphamide, doxorubicin, and an allogeneic granulocyte-macrophage colony-stimulating factor-secreting breast tumor vaccine: a chemotherapy dose-ranging factorial study of safety and immune activation. *J Clin Oncol* 2009;**27**(35):5911–8.
26. Emens LA, et al. Timed sequential treatment with cyclophosphamide, doxorubicin, and an allogeneic granulocyte-macrophage colony-stimulating factor–secreting breast tumor vaccine: a chemotherapy dose-ranging factorial study of safety and immune activation. *J Clin Oncol* 2009;**27**(35):5911–8.
27. DeGregorio M, Soe L, Wolf M. Tecemotide (L-BLP25) versus placebo after chemoradiotherapy for stage III non-small cell lung cancer (START): a randomized, double-blind, phase III trial. *J Thorac Dis* 2014;**6**(6): 571–3.
28. McNeel DG, et al. Randomized phase II trial of docetaxel with or without PSA-TRICOM vaccine in patients with castrate-resistant metastatic prostate cancer: a trial of the ECOG-ACRIN cancer research group (E1809). *Hum Vaccines Immunother* 2015;**11**(10):2469–74.
29. Heery CR, et al. Docetaxel alone or in combination with a therapeutic cancer vaccine (panvac) in patients with metastatic breast cancer: a randomized clinical trial. *JAMA Oncol* 2015;**1**(8):1087–95.
30. Kongsted P, et al. Dendritic cell vaccination in combination with docetaxel for patients with prostate cancer – a randomized phase II study. *Ann Oncol* 2016;**27**(Suppl. 6):1085P-1085P.
31. Noguchi M, et al. Mixed 20-peptide cancer vaccine in combination with docetaxel and dexamethasone for castration-resistant prostate cancer: a randomized, double-blind, placebo-controlled, phase 2 trial. *J Clin Oncol* 2018;**36**(Suppl. 6):214–214.
32. Murahashi M, et al. Phase I clinical trial of a five-peptide cancer vaccine combined with cyclophosphamide in advanced solid tumors. *Clin Immunol* 2016;**166–167**:48–58.
33. Shirahama T, et al. A randomized phase II trial of personalized peptide vaccine with low dose cyclophosphamide in biliary tract cancer. *Cancer Sci* 2017;**108**(5):838–45.
34. Kroemer G, et al. Immunogenic cell death in cancer therapy. *Annu Rev Immunol* 2013;**31**:51–72.
35. Kwilas AR, et al. In the field: exploiting the untapped potential of immunogenic modulation by radiation in combination with immunotherapy for the treatment of cancer. *Front Oncol* 2012;**2**:104.

36. Kumari A, et al. Immunomodulatory effects of radiation: what is next for cancer therapy? *Future Oncol* 2016; **12**(2):239–56.
37. Garnett-Benson C, Hodge JW, Gameiro SR. Combination regimens of radiation therapy and therapeutic cancer vaccines: mechanisms and opportunities. *Semin Radiat Oncol* 2015;**25**(1):46–53.
38. Palena C, Schlom J. Vaccines against human carcinomas: strategies to improve antitumor immune responses. *J Biomed Biotechnol* 2010;**2010**:380697.
39. Kumari A, Garnett-Benson C. Effector function of CTLs is increased by irradiated colorectal tumor cells that modulate OX-40L and 4-1BBL and is reversed following dual blockade. *BMC Res Notes* 2016;**9**:92.
40. Guo C, et al. Therapeutic cancer vaccines: past, present, and future. *Adv Cancer Res* 2013;**119**:421–75.
41. Liu R, et al. Enhancement of antitumor immunity by low-dose total body irradiation is associated with selectively decreasing the proportion and number of T regulatory cells. *Cell Mol Immunol* 2010;**7**(2): 157–62.
42. Shibamoto Y, et al. Immune-maximizing (IMAX) therapy for cancer: combination of dendritic cell vaccine and intensity-modulated radiation. *Mol Clin Oncol* 2013;**1**(4):649–54.
43. Gulley JL, et al. Combining a recombinant cancer vaccine with standard definitive radiotherapy in patients with localized prostate cancer. *Clin Cancer Res* 2005;**11**(9):3353–62.
44. Lechleider RJ, et al. Safety and immunologic response of a viral vaccine to prostate-specific antigen in combination with radiation therapy when metronomic-dose interleukin 2 is used as an adjuvant. *Clin Cancer Res* 2008;**14**(16):5284–91.
45. Kamrava M, et al. Long-term follow-up of prostate cancer patients treated with vaccine and definitive radiation therapy. *Prostate Cancer Prostatic Dis* 2012;**15**(3):289–95.
46. Garnett CT, et al. TRICOM vector based cancer vaccines. *Curr Pharm Des* 2006;**12**(3):351–61.
47. Gulley JL, et al. A pilot safety trial investigating a vector-based vaccine targeting CEA in combination with radiotherapy in patients with gastrointestinal malignancies metastatic to the liver. *Expert Opin Biol Ther* 2011;**11**(11):1409–18.
48. Kao J, et al. Targeting immune suppressing myeloid-derived suppressor cells in oncology. *Crit Rev Oncol-Hematol* 2011;**77**(1):12–9.
49. Yu H, Kortylewski M, Pardoll D. Crosstalk between cancer and immune cells: role of STAT3 in the tumour microenvironment. *Nat Rev Immunol* 2007;**7**(1):41–51.
50. Larmonier N, et al. Imatinib mesylate inhibits CD4+ CD25+ regulatory T cell activity and enhances active immunotherapy against BCR-ABL- tumors. *J Immunol* 2008;**181**(10):6955–63.
51. Pallandre JR, et al. Role of STAT3 in CD4+CD25+FOXP3+ regulatory lymphocyte generation: implications in graft-versus-host disease and antitumor immunity. *J Immunol* 2007;**179**(11):7593–604.
52. Giallongo C, et al. Myeloid derived suppressor cells in chronic myeloid leukemia. *Front Oncol* 2015;**5**:107.
53. Gill DM, et al. Association of time from definitive therapy (DT) to start of androgen deprivation therapy (ADT) for metastatic disease and survival outcomes in men with new metastatic hormone-sensitive prostate cancer (mHSPC). *J Clin Oncol* 2017;**35**(Suppl. 6):265–265.
54. Schlom J, Arlen PM, Gulley JL. Cancer vaccines: moving beyond current paradigms. *Clin Cancer Res* 2007; **13**(13):3776–82.
55. Mercader M, et al. T cell infiltration of the prostate induced by androgen withdrawal in patients with prostate cancer. *Proc Natl Acad Sci U S A* 2001;**98**(25):14565–70.
56. Gannon PO, et al. Characterization of the intra-prostatic immune cell infiltration in androgen-deprived prostate cancer patients. *J Immunol Methods* 2009;**348**(1–2):9–17.
57. Morrissey K, et al. Immunotherapy and novel combinations in oncology: current landscape, challenges, and opportunities. *Clin Transl Sci* 2016;**9**(2):89–104.
58. Ohtake J, Sasada T. Are peptide vaccines viable in combination with other cancer immunotherapies? *Future Oncol* 2017;**13**(18):1577–80.

59. Kaisho T, Akira S. Toll-like receptors as adjuvant receptors. *Biochim Biophys Acta* 2002;**1589**(1):1–13.
60. Steel JC, Waldmann TA, Morris JC. Interleukin-15 biology and its therapeutic implications in cancer. *Trends Pharmacol Sci* 2012;**33**(1):35–41.
61. Baek S, et al. Therapeutic DC vaccination with IL-2 as a consolidation therapy for ovarian cancer patients: a phase I/II trial. *Cell Mol Immunol* 2014;**12**:87.
62. Rizza P, et al. IFN-alpha as a vaccine adjuvant: recent insights into the mechanisms and perspectives for its clinical use. *Expert Rev Vaccines* 2011;**10**(4):487–98.
63. Gupta R, Emens LA. GM-CSF-secreting vaccines for solid tumors: moving forward. *Discov Med* 2010; **10**(50):52–60.
64. Goodwin TJ, Huang L. Investigation of phosphorylated adjuvants co-encapsulated with a model cancer peptide antigen for the treatment of colorectal cancer and liver metastasis. *Vaccine* 2017;**35**(19):2550–7.
65. Maisonneuve C, et al. Unleashing the potential of NOD- and Toll-like agonists as vaccine adjuvants. *Proc Natl Acad Sci U S A* 2014;**111**(34):12294–9.
66. Sharma P, Allison JP. The future of immune checkpoint therapy. *Science* 2015;**348**(6230):56–61.
67. Williams EL, et al. Immunomodulatory monoclonal antibodies combined with peptide vaccination provide potent immunotherapy in an aggressive murine neuroblastoma model. *Clin Cancer Res* 2013;**19**(13): 3545–55.
68. Karyampudi L, et al. Accumulation of memory precursor CD8 T cells in regressing tumors following combination therapy with vaccine and anti-PD-1 antibody. *Cancer Res* 2014;**74**(11):2974–85.
69. Bjoern J, et al. Safety, immune and clinical responses in metastatic melanoma patients vaccinated with a long peptide derived from indoleamine 2,3-dioxygenase in combination with ipilimumab. *Cytotherapy* 2016; **18**(8):1043–55.
70. Hodi FS, et al. Improved survival with ipilimumab in patients with metastatic melanoma. *N Engl J Med* 2010; **363**(8):711–23.
71. Cho H-I, Celis E. Optimized peptide vaccines eliciting extensive CD8 T-cell responses with therapeutic antitumor effects. *Cancer Res* 2009;**69**(23):9012–9.
72. Kumai T, et al. Optimization of peptide vaccines to induce robust antitumor CD4 T-cell responses. *Cancer Immunol Res* 2017;**5**(1):72–83.
73. Sharma RK, Yolcu ES, Shirwan H. SA-4-1BBL as a novel adjuvant for the development of therapeutic cancer vaccines. *Expert Rev Vaccines* 2014;**13**(3):387–98.
74. Jacobs JF, et al. Dendritic cell vaccination in combination with anti-CD25 monoclonal antibody treatment: a phase I/II study in metastatic melanoma patients. *Clin Cancer Res* 2010;**16**(20):5067–78.
75. Rech AJ, Vonderheide RH. Clinical use of anti-CD25 antibody daclizumab to enhance immune responses to tumor antigen vaccination by targeting regulatory T cells. *Ann N Y Acad Sci* 2009;**1174**:99–106.

CHAPTER

CONCLUDING REMARKS AND FUTURE PERSPECTIVES ON THERAPEUTIC CANCER VACCINES

14

Mahsa Keshavarz-Fathi[1,2,3], **Nima Rezaei**[3,4,5]

School of Medicine, Tehran University of Medical Sciences, Tehran, Iran[1]*; Cancer Immunology Project (CIP), Universal Scientific Education and Research Network (USERN), Tehran, Iran*[2]*; Research Center for Immunodeficiencies, Children's Medical Center, Tehran University of Medical Sciences, Tehran, Iran*[3]*; Department of Immunology, School of Medicine, Tehran University of Medical Sciences, Tehran, Iran*[4]*; Network of Immunity in Infection, Malignancy and Autoimmunity (NIIMA), Universal Scientific Education and Research Network (USERN), Tehran, Iran*[5]

CONCLUDING REMARKS

Cancer immunotherapy is rapidly moving forward to fulfil its ambition to harness the immune system and selectively destroy cancerous cells without severe harm to normal cells. Various immunotherapeutics have been promoted to interfere with defective immunity in response to malignancies. Among them, as agents with a historical background of preventing infectious diseases, vaccines had an evolutionary course following the identification of many tumor antigens, which are great targets for directing a specific immunotherapy. Vaccines had achievements in evoking immune responses against tumors and developing positive clinical responses in most early clinical trials. Immunogenicity and the safety of vaccines were reported in many clinical trials. These safe modalities showed restricted efficacy in some phase III trials. However, some positive results were obtained in the long-term follow-up, which showed their efficacy in improving patient responses and not necessarily the tumor response. In some subgroups of patients, who were randomized to the vaccine arm in the trials, significant positive clinical outcomes were yielded that were not observed in the whole sample.[1,2] These results emphasized that some patients might benefit more from vaccines and necessitate developing appropriate predictive biomarkers and applying them to select target patients for vaccination.[3] They also indicated long-term effects caused by the vaccines, which act opposite the direct cytotoxic agents and usually need sufficient time to induce a robust immune response. Because the immune system and tumor cells have a dynamic interaction and both tumor regression and progression might take place under the influence of the immune system, sufficient time is required to observe the effects of immunotherapy. An understanding of the fundamental mechanisms of the induction of an immune response and the immunoediting process, which can lead to eliminating the cancer or its escape from the immune system, is essential for designing effective therapeutics.

Vaccines for Cancer Immunotherapy. https://doi.org/10.1016/B978-0-12-814039-0.00014-X

BASIC IMMUNOLOGY OF CANCER VACCINES

Vaccines induce specific antitumor immune responses against tumor antigen(s). There are two principal steps for developing an antitumor immune response: the priming and effector phases. In the priming phase, the professional antigen presenting cells (APCs), i.e., dendritic cells (DCs), obtain tumor antigens, which are released by dying cancerous cells. If a danger signal is not present, immune tolerance occurs by ignoring the antigen. However, immunogenic cell death leads to generating danger signals that are recognized by pattern recognition receptors (PRRs) in DCs. Maturation and migration of DCs are the following step. Antigens and danger signals cause the maturation of DCs.[4] Afterward, DCs move toward draining lymph nodes to prime T cells. To generate the effector T cells, DCs transduce three signals to naive T cells. The first signal is the result of antigen presentation by the major histocompatibility complex (MHC) molecules to the T-cell receptor (TCR) of the T cells. The second signal depends on the costimulatory or coinhibitory molecules, which adjust and modulate the type of immune response against the danger signal. Subsequent to these signals, a number of cytokines, which are determinant for the type of immune response, are secreted from the DCs and T cells as the third signal.[5] $CD4^+$ and $CD8^+$ effector T cells, specific for the tumor antigen(s), are developed after the priming phase. Both $CD4^+$ and $CD8^+$ T cells are required to induce a robust immune response. To develop optimal and long-lived effector $CD8^+$ T-cell responses, $CD4^+$ T cells are necessary.[6] They are also responsible for the induction and maintenance of $CD8^+$ memory.[7]

There are some hurdles against the appropriate function of effector T cells, such as the immunosuppressive microenvironment shaped by some immune cells, e.g., myeloid-derived suppressor cells and T-regulatory cells and molecules including cytokines, chemokines, and indoleamine 2,3-dioxygenase released by cancerous cells. The physiological process of developing an antitumor immune response should be considered for intervention and by designing an effective therapeutic modality.[8–11]

Various types of vaccines have been designed to provide the first signal as well as using adjuvants to make the second and third signals. They are designed to deliver sufficient amounts of the appropriate tumor antigens to DCs.[12] Adjuvants, which provide the second and third signals, are two types: classic and novel. Old mineral salts such as alum are classically used combined with vaccines and can provide a delivery system as well. Cytokines such as granulocyte macrophage colony-stimulating factor (GM-CSF) and ligands to PRRs such as Toll-like receptors are the biologic adjuvants that may be administered as an independent immunotherapy as well. In addition to adjuvants, there are some vectors, which can act as an adjuvant. They include viral and nonviral vectors, used to deliver DNA or RNA.[12]

Whole tumor cells can be administered as a source of all tumor antigens; if they have autologous origins, they function as a personalized vaccine containing all tumor antigens specific for the patients. Usually some antigens have not been identified previously. This type can bypass challenges in identifying and selecting tumor antigens.[13] Peptide vaccines are produced based on the different methods of identifying and selecting the immunogenic peptide(s). The synthetic long peptides might be more beneficial because they can use both human leukocyte antigen (HLA) class I- and II-binding antigens to induce both $CD4^+$ and $CD8^+$ T cells, responses, and long-lasting immunity.[14] Despite these challenges, peptide vaccines are attractive owing to their simple manufacture and cost-effectiveness.[15]

The next type are the genetic vaccines composed of DNA or RNA encoding the antigen and a vector, which is usually a viral vector. Finally, APCs including DCs can be loaded *ex vivo* with the peptide, tumor cell lysate, or DNA and RNA to express the antigen and be used as vaccines.[16] There are advantages and disadvantages to each type of vaccines, which are reviewed in the related chapters. Two approved vaccines are from the whole tumor cell category, and the other one is an APC-based vaccine. Although no vaccines from peptide and genetic vaccine categories have been approved yet, they have many advantages that make them potentially useful for clinical application on a large scale.

APPROVED VACCINES

Sipuleucel-T (Provenge) is the only therapeutic cancer vaccine approved by the US Food and Drug Administration (FDA). Two personalized tumor cell vaccines were approved in Russia and Canada, but not in the United States.

Sipuleucel-T was approved by the FDA for therapy of hormone-refractory prostate cancer. It has a complex process of production and serves as an individualized vaccine. The APCs must be extracted from the peripheral blood mononuclear cells of the patient and be loaded *ex vivo* with GM-CSF as an adjuvant and prostatic acid phosphatase (PAP) as a tumor antigen.[17,18]

In the phase III trial leading to the approval of Provenge, 4.1–4.3 months' improvement in overall survival (OS) and a 22% reduction in the risk for death were observed in the vaccine arm compared with the placebo.[19] Unsurprisingly, and similar to what is expected from cancer vaccines, there was evidence of better outcomes in patients with lower tumor burden in the Provenge arm. Over 10 months' improvement in OS was reported in patients with lower prostate-specific antigen (PSA) levels in the Provenge arm.[20] There are ongoing trials assessing Provenge combined with other treatments such as ipilimumab, and radiation therapy.[21] (NCT01833208, NCT01818986).

Vitespen or Oncophage is a heat shock protein–based personalized vaccine assessed for melanoma and renal cell carcinoma. In a randomized phase III trial including two arms of Vitespen and observation for patients at high risk for renal cell carcinoma recurrence, 361 patients in the adjuvant arm and 367 in the observation arm were analyzed for recurrence-free survival. The first paper reporting Vitespen's efficacy was published in 2008, in which after 1.9 years' follow-up no statistically significant differences were seen in recurrence-free survival between groups. Subgroup analysis and an additional 17 months' observation did not differ in the results.[22] In the second report, following patients for 3 years to achieve OS as the primary end point, results were in favor of the Vitespen arm for an analysis of all intent-to-treat groups and subgroups, especially in those with lower stages and at intermediate risk of recurrence.[23] This report resulted in its approval in Russia. It was approved in April 2008 by the Russian Ministry of Public Health as adjuvant therapy for renal cell carcinoma at intermediate risk for recurrence after nephrectomy.[24]

Melacine, which received approval in Canada, is the other personalized vaccine administered to patients with melanoma. The vaccine contains tumor cell lysates and Detox-PC as adjuvant. The tumor lysates possess various melanoma antigens such as gp100 and MAGE-1–3. It was compared with chemotherapy in a phase III trial, which led to its approval.[25–27] The study demonstrated 3 months' improvement in OS in the vaccine arm (Fig. 14.1).[28]

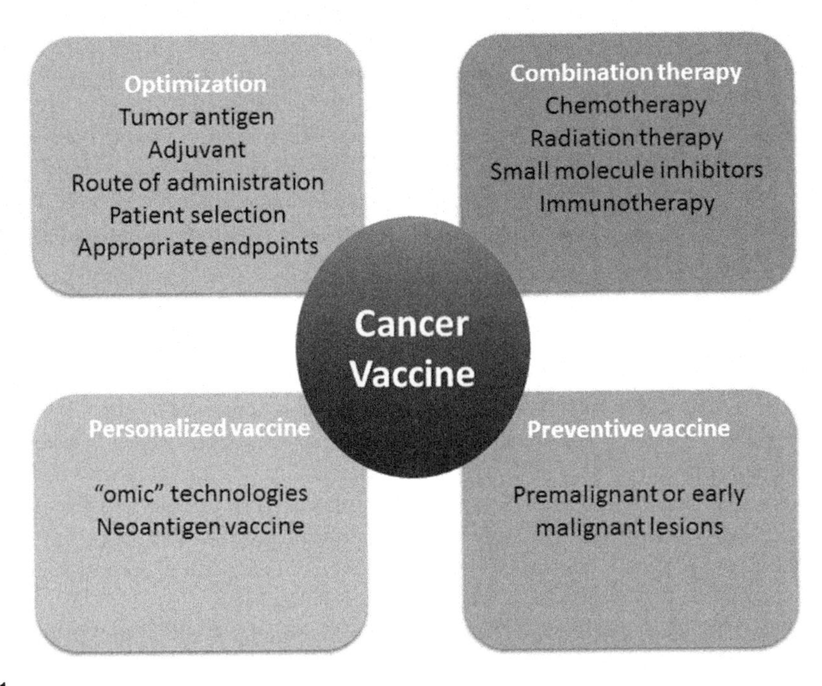

FIGURE 14.1

Considerations for clinical improvement of cancer vaccines.

NEOANTIGEN VACCINES

Progress in technologies that explore genomic data led to the development of personalized approaches in medicine such as personalized vaccines. These technologies facilitated the identification of neoantigens and the development of vaccines that target individualized tumor antigens. The positive results of two trials of neoantigen vaccines were released and provided insights into the efficacy of cancer vaccines. In one trial of a peptide vaccine containing up to 20 neoantigens in advanced melanoma at high risks for recurrence, four of six patients showed no recurrence after over 2 years' follow-up. The other two patients with recurrence responded to combination therapy with an anti-PD-1 antibody.[29,30]

In another trial of a poly-neo-epitope genetic vaccine, 13 patients with advanced melanoma were recruited and all had improvement in progression-free survival. Eight patients without any lesion prior to vaccination, were completely recurrence-free for 12-23 months. Five patients had lesions before the vaccine, two of whom had complete responses, one complete response after combination with an anti-PD-1 antibody, one partial response, and one stable disease.[31]

PREVENTIVE CANCER VACCINE

Advanced stages of cancer in patients undergoing vaccine therapy are a remarkable barrier in clinical trials of cancer vaccines. In this case, patients may be immunocompromised owing to the disease and prior treatments. Induction of an immune response providing all signals and eradicating tumors may not be achievable in this setting.[32] Therefore, the application of vaccines in earlier phases is required to prevent the progress of pre- or early malignant lesions. This concept was applied to the treatment of cervical cancer, which is induced by human papillomavirus. However, it has the potential to be used not only in viral-induced cancers but also in cancers, which are detectable in early stages, such as prostate, breast, and colorectal cancer. The development of imaging studies and screening biomarkers have a major role in the early diagnosis of lesions and even finding susceptible people, to administer preventive cancer vaccines.[33]

CONSIDERATIONS TO FULFILL AMBITIONS

Finally, some points must be taken into consideration to improve the efficacy and effectiveness of cancer vaccines, and these provide the basis for a future perspective on cancer vaccines: (1) optimization of vaccination by choosing an effective route of administration and adjuvants; (2) an appropriate vaccine schedule to boost immune responses; (3) conducting preclinical and clinical studies to figure out the mechanism of action and efficacy of novel adjuvants such as PRRs and costimulatory molecules; (4) the use of vaccines in the premalignant or early stages of the malignancy, i.e., exploiting vaccines in preventive settings; (5) development of immune criteria and predictive biomarkers that can help to select patients who benefit more from the vaccine therapy; (6) design of clinical trials appropriate for evaluating vaccine effectiveness; (7) determining surrogate end points that correlate with clinical responses and can be used to evaluate vaccines's efficacy; (8) setting end points favoring patient benefit rather than tumor benefit, i.e., using OS instead of tumor response or progression-free survival; (9) evaluation of a combination of various therapies such as conventional therapies and other immunotherapeutics with cancer vaccines; (10) application of personalized vaccines in the case of advanced cancers resistant to other treatments; and (11) improvement of methods used to identify and select tumor antigens and neoantigens.

REFERENCES

1. Giaccone G, Bazhenova L, Nemunaitis J. A phase III study of belagenpumatucel-L therapeutic tumor cell vaccine for non-small cell lung cancer [abstract LBA 2]. *2013 European Cancer Congress;* 2013. Presented September 28, 2013.
2. Butts C, et al. Updated survival analysis in patients with stage IIIB or IV non-small-cell lung cancer receiving BLP25 liposome vaccine (L-BLP25): phase IIB randomized, multicenter, open-label trial. *J Cancer Res Clin Oncol* 2011;**137**(9):1337–42.
3. Schlom J. Therapeutic cancer vaccines: current status and moving forward. *J Natl Cancer Inst* 2012;**104**(8): 599–613.
4. Galluzzi L, et al. Immunogenic cell death in cancer and infectious disease. *Nat Rev Immunol* 2017;**17**(2): 97–111.
5. Kershaw MH, Westwood JA, Darcy PK. Gene-engineered T cells for cancer therapy. *Nat Rev Cancer* 2013; **13**(8):525–41.
6. Ossendorp F, et al. Specific T helper cell requirement for optimal induction of cytotoxic T lymphocytes against major histocompatibility complex class II negative tumors. *J Exp Med* 1998;**187**(5):693–702.
7. Janssen EM, et al. CD4+ T cells are required for secondary expansion and memory in CD8+ T lymphocytes. *Nature* 2003;**421**(6925):852–6.

8. Munn DH, Mellor AL. Indoleamine 2,3 dioxygenase and metabolic control of immune responses. *Trends Immunol* 2013;**34**(3):137−43.
9. Lippitz BE. Cytokine patterns in patients with cancer: a systematic review. *Lancet Oncol* 2013;**14**(6). e218-28.
10. Gorbachev AV, Fairchild RL. Regulation of chemokine expression in the tumor microenvironment. *Crit Rev Immunol* 2014;**34**(2):103−20.
11. Nagarsheth N, Wicha MS, Zou W. Chemokines in the cancer microenvironment and their relevance in cancer immunotherapy. *Nat Rev Immunol* 2017;**17**(9):559−72.
12. Vasquez M, Tenesaca S, Berraondo P. New trends in antitumor vaccines in melanoma. *Ann Transl Med* 2017; **5**(19):384.
13. Chiang CL-L, Benencia F, Coukos G. Whole tumor antigen vaccines. *Semin Immunol* 2010;**22**(3):132−43.
14. Bijker MS, et al. Superior induction of anti-tumor CTL immunity by extended peptide vaccines involves prolonged, DC-focused antigen presentation. *Eur J Immunol* 2008;**38**(4):1033−42.
15. Keenan BP, Jaffee EM. Whole cell vaccines — past progress and future strategies. *Semin Oncol* 2012;**39**(3): 276−86.
16. Tagliamonte M, et al. Antigen-specific vaccines for cancer treatment. *Hum Vaccines Immunother* 2014; **10**(11):3332−46.
17. Patel PH, Kockler DR. Sipuleucel-T: a vaccine for metastatic, asymptomatic, androgen-independent prostate cancer. *Ann Pharmacother* 2008;**42**(1):91−8.
18. Di Lorenzo G, Ferro M, Buonerba C. Sipuleucel-T (Provenge(R)) for castration-resistant prostate cancer. *BJU Int* 2012;**110**(2 Pt 2):E99−104.
19. Kantoff PW, et al. Sipuleucel-T immunotherapy for castration-resistant prostate cancer. *N Engl J Med* 2010; **363**(5):411−22.
20. Schellhammer PF, et al. Lower baseline prostate-specific antigen is associated with a greater overall survival benefit from sipuleucel-T in the Immunotherapy for Prostate Adenocarcinoma Treatment (IMPACT) trial. *Urology* 2013;**81**(6):1297−302.
21. Schepisi G, et al. Immunotherapy for prostate cancer: where we are headed. *Int J Mol Sci* 2017;**18**(12):2627.
22. Wood C, et al. An adjuvant autologous therapeutic vaccine (HSPPC-96; vitespen) versus observation alone for patients at high risk of recurrence after nephrectomy for renal cell carcinoma: a multicentre, open-label, randomised phase III trial. *Lancet* 2008;**372**(9633):145−54.
23. Wood CG, et al. Survival update from a multicenter, randomized, phase III trial of vitespen versus observation as adjuvant therapy for renal cell carcinoma in patients at high risk of recurrence. *J Clin Oncol* 2009;**27**(15S). 3009−3009.
24. Carlson B. Research, conferences, and FDA actions. *Biotechnol Healthc* 2008;**5**(1):7−16.
25. Fintor L. Melanoma vaccine momentum spurs interest, investment. *J Natl Cancer Inst* 2000;**92**(15):1205−7.
26. Dubensky TW, Reed SG. Adjuvants for cancer vaccines. *Semin Immunol* 2010;**22**(3):155−61.
27. Jain KK. Personalized cancer vaccines. *Expet Opin Biol Ther* 2010;**10**(12):1637−47.
28. von Eschen KB, Mitchell MS. Phase III trial of melacine® melanoma theraccine versus combination chemotherapy in the treatment of stage IV melanoma: 179. *Melanoma Res* 1997;**7**:S51.
29. Bullen Love D. The potential of personalized cancer vaccines. *Oncol Times* 2017;**39**(18). p. 1,8-8.
30. Ott PA, et al. An immunogenic personal neoantigen vaccine for patients with melanoma. *Nature* 2017; **547**(7662):217−21.
31. Sahin U, et al. Personalized RNA mutanome vaccines mobilize poly-specific therapeutic immunity against cancer. *Nature* 2017;**547**(7662):222−6.
32. Gray A, et al. A paradigm shift in therapeutic vaccination of cancer patients: the need to apply therapeutic vaccination strategies in the preventive setting. *Immunol Rev* 2008;**222**:316−27.
33. Melief CJM, et al. Therapeutic cancer vaccines. *J Clin Invest* 2015;**125**(9):3401−12.

Index

'*Note*: Page numbers followed by "f" indicate figures and "t" indicate tables.'

Q

R

W

Printed in the United States
By Bookmasters